We are Big Data

Sander Klous · Nart Wielaard

We are Big Data

The Future of the Information Society

ATLANTIS
PRESS

Sander Klous
Informatics Institute
University of Amsterdam
Amsterdam, North Holland
The Netherlands

and

Management Consulting
KPMG
Amstelveen, North Holland
The Netherlands

Nart Wielaard
Nart BV
Haarlem, North Holland
The Netherlands

ISBN 978-94-6239-182-6   ISBN 978-94-6239-183-3   (eBook)
DOI 10.2991/978-94-6239-183-3

Library of Congress Control Number: 2016937342

Printed on acid-free paper

Art becomes art when you can see the complete picture.
Data acquires value when you can understand the context.

---

[1]What is this? It will become clear as you read this book.

Foreword

Big Data matters—a lot. But in a more subtle and fundamental way than is often portrayed today. Marketers from consulting and IT companies tout the virtues of Big Data with deafening volume, limited variety, and increasing velocity. Their mantra is that with new IT and a consulting partner lined up, any company can turn itself into a Big Data superpower. That is patently wrong—and the marketers' behavior is as predictable as it is irresponsible.

By the same token, commentators and self-styled cyber-experts paint dire monochromatic pictures of a Big Data future in which everything human will have become eliminated. Listening to some of them, it seems that with Big Data, we'll turn our computers into weapons of mass destruction. Such doomsday predictions may sell books but, just like the Big Data hype do little to improve our understanding of what Big Data actually is—and how fundamentally it will change our economy and our society.

Big Data is, at its very core, nothing more than process, a unique mechanism of how we humans make sense of the reality around us and, based on this understanding, make predictions about our future that are far more accurate than the tea-leaf reading of the past.

For all of human history, we have made sense of the world that surrounds us by watching it and thinking about it—by theorizing about how reality comes together. And we gathered data about the world to prove (or disprove) our theories. Over time, we realized the need to see the world in rational terms, and to engage in understanding it in a methodical fashion. This is the seedling that bloomed in the Age of Enlightenment and which was nurtured among others by Spinoza, one of Holland's intellectual heroes. Eventually this yielded the "scientific method" and the great progress in understanding and utilizing the world in the 19th and 20th centuries. As a result, many more people have been living longer and better.

But until recently, the collection, analysis, and storage of data was time-consuming and costly. Thus, we collected as little data as possible to answer the questions we had. In fact, in the age of small data, our institutions and processes the very way we gained insights from data was premised on data scarcity. So we used samples of data rather than all of it—for everything from quality control in manufacturing to polling voters before elections. That shortcut worked if done correctly (which was not always the case), but it lacked details. So we could rarely look deeper into something that turned out to be interesting, because we simply did not have enough data.

More troubling, with such an approach we could only answer questions we already had at the outset, not use the data effectively to pose new questions that might lead to truly novel insights. We also privileged data quality, striving even at high cost to improve the accuracy of the little data we collected.

Today, many of the barriers to collecting, analyzing, and storing large amounts of data have been greatly reduced. This makes it vastly easier to collect all, or close to all, the data and capture the complexity of our world rather than settle for just a small sample of reality. In many cases we can also choose to collect more, but messier, data rather than gather very small amounts.

As a result, we can now look at the world in unprecedented detail and Big Data analysis yields new questions that point us toward valuable insights we otherwise would not have. This is how Google is attempting to predict the spread of the flu by analyzing the billions of search requests it receives. This is how Rolls Royce can foresee when a part in one of its airplane engines is likely to break before it actually does, which not only saves money but also saves lives.

And this is how startup companies like INRIX are able to direct their customers around traffic jams on their daily commutes as well as advise local governments on emerging traffic patterns. This makes IBM's Watson computer win against the best humans in *Jeopardy*, and help doctors in their diagnoses. At the University of Ontario they have even found a way to save the lives of premature babies harnessing Big Data. In the Big Data age, more and messy data has taken over from clean, but small, data.

But there is a third important shift in how we make sense of the world. We humans prefer to understand the world as a sequence of causes and effects. And so we cannot help but look for causes of everything we notice—unexpected noises, a car that won't start, a shopper's decision to buy, the way we feel. But as it turns out, identifying causes is hard. So, as Nobel laureate Daniel Kahnemann has shown, we often tend to intuit causes without conclusive evidence. That comforts us and it gives us a false sense that we understand the world.

Much of Big Data analysis cannot tease out causes from analyzing effects. It regularly cannot tell us the "Why", only "What". But, amazingly, that is often good enough to improve our decision-making. It helps Google translate dozens of languages on the fly and accounts for about a third of Amazon's sales. It even lets cars drive themselves.

This of course does not mean that our lifelong quest for causes suddenly should come to an end. Rather, it helps us realize that before we can run, we need to be able to walk. With Big Data we have a powerful tool to do just that. If applied appropriately, Big Data analysis will discover the "What", thus acting as an invaluable filter for subsequent thorough causal analyses.

This helps us appreciate that the world we live in is far more complex, but also far more interesting and thought-provoking than we thought. As we realize that the data we can now gather holds so much latent value which we can uncover through Big Data, businesses will rethink how they operate, reshaping our economy and helping our society evolve.

In this smart and insightful book, written by experts who truly understand and grasp Big Data, you will gain a new appreciation for Big Data and how we can harness it for the good. It offers a passage to an exciting journey that can yield amazing insights and produce significant value. Importantly, you will also hear how Big Data's power can be channeled so that we can benefit from it and avoid suffering from some of its negative consequences.

Big Data matters—a lot. Make it matter for you!

Viktor Mayer-Schönberger
Co-author of Big Data: A Revolution That Will
Transform How We Live, Work, and Think

Contents

Introduction: Anything is Possible

Expecting the extreme is the new norm

The iPad Generation

YouTube[1]. We see a baby—so young that she can only coo—in front of several glossy magazines. She is turning pages and wants to zoom in or click through. The usual swiping, pinching and tapping— she has obviously already mastered on a tablet at this young age—isn't working. Disappointed, she throws the magazines aside. This was filmed by her father, Jean-Louis Constanza, then-CEO of Orange Vallée, who sums it up beautifully: "In the eyes of my one-year-old daughter, a magazine is a broken iPad. That will remain the case throughout her entire life. Steve Jobs programmed part of her operating system."

New technology has made many incredible things possible. It is now part of our lives in a way we couldn't even have dreamed of 20 years ago. Okay, no more clichés...

Don't worry; this book isn't (really) about technology.

We could throw around some impressive numbers about the phenomenal calculative capacity required for the successful search for the Higgs boson particle—the 'God particle'—at CERN, the European organization in Geneva that conducts fundamental research into elementary particles and is therefore in the Champions League of data analysis. (One of the authors, Sander, was involved with the Atlas experimentat CERN, which analyzed the data obtained by the particle accelerator.) We could explain that new open-source products are the basis of numerous new groundbreaking data applications. We could attempt to unravel the algorithms that allow Google to return such good results for your searches.

The New Information Society

Each of these is an interesting technological subject, but we only discuss them briefly in this book. We prefer to dwell on the impact of this kind of technology on society and, more specifically, how this technology, related to data, data processing, information, and communications, can improve our society. A new world is rapidly emerging, a new world that, in this book, we call the new information society. It is a world in which everything is measurable and in which people, and almost every device you can think of, are connected 24/7 through the Internet. That network of connections and sensors generates a phenomenal amount of data and offers fascinating new possibilities that, together, are often called Big Data. Without a doubt, these possibilities are vast and they will be exploited. They're already being exploited! We are in the middle of this change, even if we are not always aware of it.

Awareness of this change is spreading through businesses and governments who are asking themselves how to respond to these developments. This goes beyond launching new products or services. It's about how to respond to a totally different world. In this respect, the term 'age of disruption' is increasingly being used. The term describes the disruptive effect that rapid and successive technological breakthroughs are having on virtually all aspects of society, which is definitely challenging, but also offers enormous opportunities.

In macroeconomic terms, there is reason for great optimism. We are in the middle of a classic economic phenomenon: 'the creative destruction'.[2] This means that we can only achieve innovation once the old order has been dismantled, which is already happening in almost all sectors. Big Data is acting as a growth hormone for this creative destruction. The growth is undeniable. All our chatter on social media, online music and video streaming, our e-mails, online shopping transactions and so on result in barely comprehensible amounts of data traffic. And that mountain of data is growing rapidly now that the 'Internet of Things' is linking billions of devices which were previously 'offline'—TVs, refrigerators, security devices, thermostats, smoke detectors—all of which now produce and share data.

It's tempting to start throwing impressive figures around here, but that's not the purpose of this book. One figure is actually enough: in the last two years, the amount of data recorded was 10 times greater than the amount previously recorded over the entire history of humanity.[3] This seems impressive, but what's really impressive is what you can do with that data. That is what we discuss in this book.

The amount of data produced is growing exponentially, to astronomical levels, yet the word 'big' in Big Data is actually slightly misleading. Many new applications are not about editing or interpreting enormous amounts of data—the so-called 'big, messy data'—but much more about the smart combination of data.

An example is the rise of the 'Quantified Self' phenomenon,[4] the trend where an increasing number of people are measuring numerous aspects of their lives. They want to gain insights into what they're eating, how much they're moving, how deeply they're sleeping, what happens to their heart rates when they exercise. They themselves determine the level of detail they want to measure. For example, they can record how much coffee they drink or how much chocolate they eat. Through various apps on smartphones or other devices, they can simply record information on their own behavior, health and lifestyle and can then use this data to motivate themselves, for example, to reach the next level of fitness. Semantically, this has little to do with *Big* Data. However, it forms part of what society considers Big Data.

Big Data in This Book

We have chosen to use Big Data as an umbrella term. Big Data, in this book, includes all the new opportunities, possibilities, techniques, and threats associated with the fact that we can now deal with data differently or, in other words, the positive and negative aspects of the 'datafication' of society, including social facets such as privacy and social influences. This includes the extensive monitoring of personal behavior by intelligence agencies, commercial precision-bombing of customers, resolving traffic congestion, stopping epidemics and connecting a refrigerator or thermostat to the Internet. An important part of Big Data is data analysis: searching for patterns in the large amounts of data to enable relevant conclusions to be drawn.

Relationship with Technology

We believe that Big Data can be extremely valuable to the world. We are already experiencing some of the amazing benefits every day, as technology has touched almost every aspect of our lives. We are all familiar with the technology used to communicate with friends and family all over the world. We can send them text messages on smartphones, play online games with them on our tablets or keep in touch via social media. We can also take care of practicalities through the wonders of technology, like submitting bills from our physiotherapists straight to our health insurers over the Internet.

We store photos and videos in the cloud—which is very handy when we want to share them—and we use the cloud to avoid holding all of our business data on our own hard drives. Keeping things online is easier and reduces the risk of us losing them. We arrange our banking with an app. When we need information about our cocker spaniel's sudden symptoms, we find it on the Internet in an instant. When we're arguing in a bar about whether Peter O'Toole did or didn't win an Oscar, it takes 10 seconds to find the answer with a smartphone. As technology entrepeneur Peter Hinssen said in one of his books: "Digital is the new normal in everything we do. Digital living is not only something we do completely routinely, we have also become addicted to it and our expectations about the possibilities of this digital life have increased tremendously."[5]

Expectations

Where will it end? What will digital life look like in the future? Are we all going to be wearing Google Glass—or its decendants—as a replacement for our smartphones? Or will the smart watch break through? Perhaps. Will employment decrease by half over the next 50 years because more and more tasks, such as writing a book like this, will be taken over by computers? Maybe. Will we be able to extend our average life-span by 20 years, in part because we have the computing power to map the human genome? It's possible. But the honest answer is that nobody can say with any certainty how the use of technology will evolve.

We have no doubts that technology—and hence Big Data—will infiltrate every aspect of our lives even more in the years to come. Not only because it's impossible to avoid, but also because we actively want it. Because we don't want to live without the convenience, comfort, or added value that technology brings.

Technology creates expectations, like the toddler who expected to use a paper magazine as a tablet. We should be able to do anything; have an app for everything. The uptime of all services needs to be 100 %. Companies will need to provide tailor-made products, preferably at zero cost. Moreover, it will all need to be fun and sustainable.

All of this makes it easy to summarize the new challenge facing companies, governments, and other organizations: *the impossible needs to be available immediately; delivery within two weeks is permissible for miracles.*

Possibilities

Big Data is more than just a new technology or concept. It offers opportunities to all sectors and industries for us to organize ourselves differently, to make progress and to do things that were simply not possible until recently. Many people associate the term 'Big Data' with companies wanting to sell their customers even more stuff by learning everything there is to know about them. However, that's only one side of the coin, a side that we will definitely address in this book. We also want to show that Big Data can solve societal problems and improve our lives. We can use it to save human lives. We can improve maintenance and plan more cost-effectively. We can improve traffic safety and reduce congestion. We can increase agricultural yields. We can provide a better quality of life for the elderly. We can make the world more sustainable. We can revolutionize medical diagnostics and treatments. We can increase the security of payments. We can even improve a team's sporting performance.

Structure

We begin this book with an overview of what we can do with Big Data and elaborate on both the benefits and the drawbacks. We discuss how we, as citizens and consumers, have extreme expectations, but we also feel resistance against the appetite of organizations for our personal data and we fear for our privacy. In doing so, we reach the conclusion that we cannot deal with Big Data in isolation, because it is fundamentally linked to social progress.

That's why in the second part of the book we discuss the new information society that is emerging, of which Big Data is an essential part. We show that individuals are increasingly demanding that organizations give back to society rather than acting only for gain, that we will become ever more powerful as a collective of individuals and that, as a collective, we can have a significant influence on the actions of organizations. In this new world, we must find a way to embed Big Data successfully into our society.

Broadening our scope to take in the information society as a whole leads, in the third part of the book, to a discussion of how Big Data can be successfully embedded into society. Issues that are addressed include ethical principles, education and the role of government.

Big Data can—in a controlled manner—contribute to a better society. That's exactly why we wanted to write this book. We want to show that companies and governments still don't know how to deal with Big Data. The subject is dogged by misunderstanding and ignorance. Big Data projects are often fragmented and unstructured. This leads to unsuccessful projects, significant disappointment, and resistance. This must change. Throughout this book, we aim to show how things could be different. In the following chapter, we first look at the (potential) advantages that Big Data has to offer us.

Chapter 1
Big, Bigger, Biggest Data

New opportunities, made possible by an abundance of data

Will It Ever Be Possible to Predict Crime?

Actor Tom Cruise storms into a house in the suburbs of Washington and runs straight to the bedroom, just before the resident tries to stab his wife—whom he found in bed with another man—in the chest. Cruise states his name is John Anderton, chief of a police unit focused on crime prevention, and arrests the man for the murder he was about to commit. This opening scene of the movie Minority Report suggests that in the year 2054, predictions will be so exact that people will be arrested before they commit a crime.

Will it ever be possible to predict crimes through data analysis? Should we want to? The reality is that several American cities are already successfully using 'predictive surveillance'. Based on the analysis of large amounts of data from various sources, algorithms determine which streets require additional attention from police officers. According to Predpol, one of the key suppliers of predictive policing software, the effect will be more arrests, but also lower crime rates. An experiment conducted in a town in Kent, England showed that number crunching is significantly better at predicting where a crime will occur than the conventional approach. Data analysis made a correct prediction in 8.5 percent of cases, whereas predictions made by experienced professionals scored no better than 5 percent. In an earlier experiment in Los

© Atlantis Press and the author(s) 2016
S. Klous and N. Wielaard, *We are Big Data*,
DOI 10.2991/978-94-6239-183-3_1

Angeles, California, these scores were 6 percent and 3 percent. *The Economist*, a British weekly magazine, had a pointed title for an article about this: "Don't even think about it".[1] However, some experts have expressed criticism regarding the expediency and possible side effects of such measures. Namely, an individual's right to privacy could lose out to society's need for safety.[2]

In another example, again from the US: Baltimore's Parole Board uses risk profiles that indicate the level of risk of a detainee becoming the perpetrator or victim of a murder after being released from custody.[3] This risk calculation forms the basis for deciding what level of monitoring is required for the detainee about to be released.

These are all examples that show how we can increase societal safety by analyzing data in a smart manner. Restraint is sometimes required and policy transparency is a must. For example, in the case of releasing detainees based on data analyses, it is important that the authorities clearly show the public how decisions are made.

Not New

Of course, Big Data is not completely new. For decades, research institutions and companies have gathered large amounts of data in order to generate new information. Stock exchange traders use models that draw on extensive data streams from various sources for the early detection of trading risks and opportunities. Intelligence agencies search for patterns in data in order to prevent potential attacks. Insurers use data to estimate the risk levels of individual customers and entire portfolios. The unifying theme in this respect has been the same for decades: we search for correlations, early indicators and cause-and-effect relationships between phenomena, persons and events and make decisions based on the results.

However, since the advent of the Internet, things have changed. The application of data analysis has broadened and deepened and has entered our daily lives. This is further accelerated through the rise of the 'Internet of Things' in which all kinds of appliances, big and small, are connected both with each other and with us.

How and why (the use of) data analysis is changing is discussed below.

Exponential Growth of Information Technology Possibilities

Gordon Moore, one of the founders of Intel, predicted in 1965 that the number of components in a dense integrated circuit would double every two years due to technological progress (Moore's law). In simple terms, every two years, twice the computational power could be bought for the same amount of money or the same computational power for half the amount. Although experts have been expecting for years that, as the limits of fundamental physics (such as the number of atoms per circuit) are reached, the prediction would cease to be true,[4] it still applies.

Comparable laws apply to network and storage capacity. In this respect, it is interesting to note that storage capacity doubles more quickly—within 12 months—than computational power—within 18 months. Since, generally speaking, we actually use that storage capacity, the conclusion is that increasingly less computational power is available per byte of stored data. In other words, data is becoming exponentially unanalyzable[5] with traditional methods, forcing a rethink on how we exploit the vast amounts of data we are producing.

As well as the exponentially increasing technical possibilities, another interesting shift is occurring. Until recently, research and progress in data analysis were mainly achieved by scientific institutions; however, nowadays the big breakthroughs are increasingly being achieved as a result of major advances by the business world. Companies such as Google, Facebook, Yahoo!, Apple, and Amazon are playing an important role in this respect. This is on top of the technological breakthroughs that were already the domain of commercial entities such as Intel, IBM and Cisco, and it is driving an accelerated learning and development process. A clear sign of this cross-pollination is the announcement by CERN[6] that it is to use Google technology, the Google toolchain, to manage the data centers responsible for the analysis of the impressive amounts of data generated by the Swiss particle accelerator.

Although the term 'Big Data' suggests that an increase in storage capacity and computational force is the key motivating factor, in reality it is the increasing capacity of networks that is driving the revolution. Twenty years ago, the possibilities in respect of computer systems were limited: there were limits to the amounts of data you could transfer and the way you could transfer data between systems. Now, an almost limitless number of sources can be reached with very little effort. Applications on devices like smart phones—such as those for streaming music—are possible because of this development.

Don't Decide Based on What We Say, but on What We Do

Traditional market research is slowly disappearing due to Big Data in light of the new ways organizations have of getting much closer to groups of customers or stakeholders. Companies no longer have to ask what people think or perceive (i.e. through market research) and just hope they will tell the truth. Instead, they can just track us. Where do we go? What do we buy? What music do we listen to? What movies do we watch? Whom do we call? Now that companies can learn the answers to these questions, they can truly customize their products and offer us new, targeted services.

Some large companies in the tech field, with Google, Apple, and Samsung at the fore, are very aware of this. They would like us to use their products—cameras, phones, televisions—not only because they make money selling us their hardware, but also because they can stick another 'sensor' onto us. These sensors bring them ever-greater insights into our behavior, and this is most-likely where their future business lies. The 'Internet of Things' will explode according to Gartner (an American information technology research and advisory firm) which predicts that 26 billion appliances will be connected via the Internet by 2020.[7] (A significant portion of these appliances will be in or on our bodies and will measure our health.)

In any case, actions speak louder than words. Facts about our behavior are much more important than our opinions, if only because a large part of our behavior is driven by habits and routines and definitely not always by thoughtful decisions. Retail companies are becoming more aware of this and are increasingly applying mathematical insights to tap into these patterns. Mathematics has even entered the boardroom. The benefit is that key decisions are increasingly based on logic (supported by math) and less frequently on emotions or the (unsupported) opinion of a dominant board member.

Big Data is therefore ushering us into a new era, an era in which we measure and arrive at decisions differently. In an interview with The New York Times[8], Erik Brynjolfsson, a researcher at the prestigious Massachusetts Institute of Technology (MIT), compared Big Data to the pioneering work of Antoni van Leeuwenhoek. Just as his invention of the microscope opened the eyes of many other scientists in the 17th century, Big Data is also ushering in a new way for us to observe the world.

Quantity Over Quality

As we are increasingly basing our decisions on data, it is essential that we do this based on reliable data. Right? Strangely, however, this is not always the case; there are other options. For example, in many cases, it is possible to increase the level of reliability by adding more data, which fits completely with one of Google's mantras: quantity over quality. The company's reasoning is simple: all errors in data can be resolved by adding more data. Organizations such as Google therefore generally do not focus on increasing the quality of the original data—they do not even have that option—but focus instead on accumulating larger volumes of data that are combined, using specific algorithms to test them for reliability. New data sources are added, often without regard to costs or effort, like a map of all the Wi-Fi access points in the world to be able to get more accurate localizations.

We have to learn to think long and hard about the demands we make of the information that forms the basis for our decisions. Sometimes we need outcomes that are completely accurate and precise; sometimes an indication will suffice. An example to clarify this: imagine you buy a basketball in a store for €30. You need this to be the exact amount taken from your bank account. A marketing director that receives management information on his or her product sales, however, will not be interested in whether the monthly turnover is €120,951.46 or €117,345.98. What he or she is interested in is global sales trends, bestsellers, regional differences, etc. The marketing director will accept a large margin of error because his or her decisions do not depend on the precision of the information.

Even if the data is pretty unreliable, in many situations it is possible to bring the reliability up to the necessary level by adding more data which will smooth out any errors. This can also be clarified by examples. First of all, imagine that we want to determine the average height of children in a classroom. We can take a random sample and determine the height of a few children.

However, the accuracy of the average is not very high. Should we require a more accurate average, we need only include a larger number of children in the random sample. The measuring method and measuring equipment can stay the same; we just take more measurements (i.e. gather more data).

Another example can be found in a very different environment: the Large Hadron Collider particle accelerator in CERN, near Geneva, Switzerland. In this accelerator—a ring with a circumference of 27 km with huge detectors that are taller than Big Ben in London, England—numerous protons collided in recent years at almost the speed of light. During these collisions, where protons break into various particles, traces of the Higgs boson, which was undetectable prior to 2013, were found. Some years ago, the scientists at CERN found that their measurements were influenced by the position of the moon.[9] When the moon is above Geneva, the water level in the lake is a fraction higher; this puts pressure on the sides of the lake, stretching the accelerator ring below ground a little bit. The result is that the energies of the revolving particles also change somewhat, resulting in different outcomes. Without the external data on the position of the moon, it would be impossible to understand these effects. Here also, albeit in a very different manner from the height-related example above, adding data can result in increased reliability and improved understanding.

Adding additional data can thus be an important tool to arrive at results that are sufficiently accurate and reliable for our intended purpose. Many challenges regarding reliability and accuracy can be solved in this way if we can let go of the presumption that we need to find an exact outcome—which was never possible anyway—and instead limit ourselves to answering a question in a certain context. This principle also applies to the fundamental change in our media consumption: 20 years ago, we read a newspaper or other source of information and in the absence of anything better, we believed the information it contained to be 'true'. Now we face the challenge of gleaning our own truth from an overcrowded media world. This requires combining sources and making a probability analysis.

New Possibilities

It is clear that Big Data has numerous potential applications. However, there is a certain skepticism regarding how meaningful they are. The common question in this respect is whether Big Data really offers significant progress. Let us list some examples in order to answer that question.

Saving human lives

eCall is a system through which a car itself, without human intervention, alerts the emergency services when an accident occurs, communicates its location, and opens an audio connection between the car and the emergency services. For example, the car manufacturer BMW has already taken the decision to include this system in all new cars sold in the Netherlands. The EU has made this system mandatory for all new cars sold from 31 March 2018. It claims that the system will save 2500 human lives annually[10] by making better use of the period known as the 'golden hour' after a traffic accident, in which rapid and appropriate action means the difference between life and death.

Improving wind turbine maintenance scheduling and making it more cost-efficient

Wind turbine maintenance is still rather conventional and mainly comprises visual inspections and the periodic lubrication or replacement of various parts. One of the problems with wind turbines is that they are not sufficiently available due to unexpected shut down and subsequent unscheduled—and expensive—maintenance. A consortium of 18 expert parties has shown that wind turbine maintenance can be made smarter and more effective through the use of sensors—sensors that measure corrosion on the turbines, sensors that record the vibrations and movements of the tower and rotor blades, sensors that measure the oil level and temperature and sensors that monitor the weather conditions. The effect, according to the consortium, is an efficiency gain of 10 percent,[11] which holds the promise that wind turbines may at some point in the future be exploited without the need for government subsidies.

Greater road traffic safety

For years now, Google has been developing a car that can maneuver through traffic completely independently. A central pillar of this development is the interpretation of enormous amounts of data about the car's environment. The test results are promising and Google claims that the computer using this data is capable of safer road behavior than human drivers. Meanwhile, the entire automotive industry has been given a wake-up call and there are numerous promising emerging developments. It is likely that the completely autonomous self-driving car is still in the (not so distant) future. Meanwhile, 'highly automated driving', in which the car assists the driver using data from its environment, has already arrived.[12]

Better harvests

The start-up insurer The Climate Corporation is offering farmers a new type of insurance, based on machine learning algorithms and Big Data analysis. The Climate Corporation analyzes the weather conditions (and expectations) for each acre of land based on large amounts of data and then combines this data with data on harvests and soil composition. The company not only offers insurance against the weather in a way that other insurers cannot offer, but has also changed the essence of an insurer's role, with the insurer now advising farmers on how to optimize their yields and at the same time minimize their insurance claims. The agri-giant Monsanto has acquired the startup for $930 million, putting 'precision agriculture' high on its agenda.[13]

Smarter investing

The founders of the Correlation investment fund take quantitative analysis of investment one step further. They invest purely based on cold, hard data analysis and have no interest in a place on the supervisory boards of the companies in which they invest. Even

though analytics always played a major role in investment decisions, trusting only on analytics is a new frontier.[14]

Better medical treatments

Some experts say that it is only a matter of time before cancer
will no longer be a deadly disease, but instead a chronic condition.[15] One of the key factors behind this progress is DNA mapping. Scientists accomplished this for the first time in 2001 after
many years of painstaking work and it can now be done for a few
hundred euros by commercial labs. The Cancer Genome Atlas, a
collaboration between the National Cancer Institute (NCI) and the
National Human Genome Research Institute (NHGRI) in the US,
has set up a databank of DNA profiles of 11,000 patients. In order
to advance cancer research even further, exponentially more data
is required and therefore the National Cancer Institute has begun
sharing that data, anonymized, in the cloud. This is only one
example of how Big Data can be useful in medical research; in
Chap. 7, we address the great potential of data analysis in health
care.

Improving team performance

In the movie *Moneyball* (2011),[16] Brad Pitt, playing a coach, is
tasked with boosting the performance of his baseball team on a
very limited budget. After meeting a young economist, he decides
not to buy players based on intuition and hunches, but to have
statistics do the job. He goes on to acquire various underap-
preciated players for the club. This type of approach is becoming more accepted in top-level sports. Even in the conservative
world of soccer, this statistical approach is gaining ground. The
Danish soccer club, FC Midtjylland, is replacing irrational, subjec-
tive, and emotional decisions with a scientific approach based
on data analysis, allowing the small club to compete with clubs
with significantly bigger budgets. In the 2014–2015 season, the
club became the Danish league champion for the first time in its
history. As the club's owner knows, this in itself is only anecdo-
tal evidence that the approach works. He claims that the team

hardly even looks at its place in the rankings now: "Our mathematical model is always preferred above our position in the rankings, when it relates to assessing our performance."[17]

We feel that these examples make it very clear: almost everything is driven by data and this offers numerous advantages. The former European Digital Agenda Commissioner Neelie Kroes summed the previous examples up in one tweet: "Big Data is at the heart of solving most of our unsolved problems—from cancer to climate change. We must build trust in it."[18]

Re-using Data

What can companies do to respond to the opportunities offered by Big Data? One of the (relatively painless) possibilities is considering whether the data they already have can be used for different applications. With a little creativity, there are often opportunities for various new applications. Large accountancy firms, for example, are sitting on an enormous pile of data related to businesses—their customers—potentially affording them the opportunity of becoming an alternative to central statistics agencies. Telecom companies have actual location data on their customers and could potentially offer better (predictive) traffic information than anyone else.

During one of his master classes, Viktor Mayer-Schönberger, Professor of Internet Governance and Regulation at Oxford University, gave a number of examples of what he calls re-using data. One of those examples is the German airline Lufthansa gathering data on board its aircraft for use as a valuable supplement to weather predictions.[19] This allowed the company to formulate much more detailed local predictions. Such local predictions meet consumers' demands more closely: we don't care about knowing whether it will rain 'in the middle of the state' tomorrow; we want to know if we're likely to get soaked on our favorite jogging route in the local park. Another example is the

American company INRIX, which specializes in gathering and processing traffic-related information based on location data from over 250 million real-time anonymous mobile phones, connected cars, trucks, delivery vans and other fleet vehicles equipped with GPS locator devices. It then offers its users predictions regarding possible delays via an app. However, the company is also experimenting with alternative uses for its data. Since it has built up critical user mass, it can also map traffic flows in the parking lots of large retailers such as Wal-Mart.[20] Financial analysts can then predict the turnover of such chains based on these models. The renowned financial institution UBS does something similar—also focused on Wal-Mart's parking lots—but based on information about traffic in the parking lots gathered by satellite.[21] It appears likely that meaningful applications for every company's data could be found by someone capable of looking at the data differently.

Understanding Human Behavior

The potential of Big Data extends beyond just developing new applications however; it also makes it possible to interpret human behavior.

This offers organizations the potential to improve management (and, more specifically, the decisions made by management). Up to now, management has mainly been based on the estimates and judgments we make based on our experience and knowledge. We interpret the past and believe that this provides us with a good basis on which to make decisions for the future. However, in many cases, this is an incorrect—and sometimes even a disastrous—approach. Our experience can easily fool us, because we have created a personal and colored perception of our experiences, or because we are strongly influenced by our environment (groupthink). In an era in which we have an abundance of data and smart algorithms, we can do better, and Andrew McAfee—one of the authors of *The Second Machine Age*—spells this out in an article titled 'Big Data's Biggest Challenge? Convincing

People NOT to Trust Their Judgment'. He refers to, among others, psychologist Paul Meehl who started the research into human 'experts' versus algorithms almost 60 years ago. At the end of his career, Meehl summarized, "There is no controversy in social science which shows such a large body of qualitatively diverse studies coming out so uniformly in the same direction as this one." His conclusion was that algorithms were clearly better than people at making decisions.

McAfee's conclusion is also clear: "Most of the people making decisions today believe they're pretty good at it, certainly better than a soulless and stripped-down algorithm, and they also believe that taking away much of their decision-making authority will reduce their power and their value. The first of these two perceptions is clearly wrong; the second a lot less so."

In order to drastically improve management therefore, we have to let go of some of the power.

This is likely to be an enormous challenge: people have a deep-rooted tendency to make decisions based on a combination of their experiences and emotions. With advanced data analysis becoming available, our natural reflex is to consider the outcome of these models in our decisions while still believing that we as humans make better decisions. This is not the case. In his book *Super Crunchers*, Ian Ayres wrote, "Instead of having the statistics as a servant to expert choice, the expert becomes a servant of the statistical machine." Today a person's mind supported by the artificial intelligence of a machine is better than the artificial intelligence alone, but it is estimated that this will no longer be the case in five years' time. The artificial intelligence will then be better without the 'help' of a human.

The challenge for management goes beyond that, however. The entire management profession will have to be reinvented in light of the potential for mapping exactly how people collaborate, communicate, and arrive at ideas. Here also, hunches will lose out to the surgical precision of Big Data.

To learn more about this, we have to study the work of the Big Data pioneer Sandy Pentland. His work in the domain of social physics offers an entirely different approach to improving business in these challenging times. Pentland argues that our social world, no less than the material world, operates according to rules. There are "statistical regularities within human movement and communication", he writes in his book *Social Physics: How Good Ideas Spread—The Lessons from a New Science*.

The core tenet of social physics is that Big Data—or, more specifically, our expanded ability to gather behavioral data—allows for "a causal theory of social structure" and "a mathematical explanation for why society reacts as it does." Pentland—who has carried out many interesting experiments to support his theory—offers us a new way to understand human behavior by analyzing Big Data. This will give us even more insight into all aspects of human life and how ideas evolve and spread. For businesses, it unlocks all sorts of possibilities and offers a path toward building more cooperative, productive and creative organizations.

This is precisely what almost every company needs right now to survive in a time of massive change. For starters, companies can improve their management practices, building on the ideas of social physics. Pentland argues that people tend to view management as an 'art' nowadays. Social physics takes this art and turns it into a science by allowing decision-makers to qualify things, find the patterns, visualize, and then manage them. In other words, it's adding discipline to the art.

More specifically, understanding social physics may also help businesses in a variety of their business processes. In an interview for a magazine published by the company KPMG,[22] Pentland touched upon some examples. A particularly interesting one was, "We looked at a large body of bank data from developing countries and found that we could achieve almost 50 percent more accurate credit scoring than the bank could because we were looking at the social physics—the behavior of the people—rather than traditional bank data such as age, income or repayment history."

Big Data will therefore not only help us to build all kinds of applications to do great things using data. The promise for the near future is that it will also help us to better understand human behavior—a fascinating outlook with far-reaching consequences.

Big Data's Drawbacks

Big Data clearly offers a lot of potential to improve our society and our living conditions. At the same time, we realize all that glitters is not gold. If businesses and governments fail to deal with this phenomenon properly, they may experience a backlash as a result of the dangers and risks associated with Big Data. This would, in turn, increase people's resistance to embracing the potential that Big Data offers. In order to understand this more clearly, we discuss the dark side of Big Data in the next chapter.

Chapter 2
Remove Fear and Conquer Resistance

Businesses need to emphasize the societal value of Big Data

People Are Being 'Milked'

Your personal data is valuable. Companies like Facebook and Google are making money from something that's not really theirs—our data. According to some, we are being 'milked', like in the sci-fi movie The Matrix. In this 1999 movie, people's energy was used to support a post-apocalyptic robot world. Lateral thinker Jaron Lanier sketches similar scenarios but suggests an alternative view. In his 2013 book Who Owns the Future?, he argues that we shouldn't see computers as passive tools, because in doing so we fail to understand how computers and human beings interact. Moreover, he wants people to reclaim their own economic destiny by creating a society that values the work of all industries and not just those with the fastest computer networks.

Whether we like it or not, data is playing an increasingly prominent role in our lives. And for some time now, this has been happening not just in communications, but also in a whole host of different areas. Using data (networks) means, for example, that traffic lights can be synchronized to traffic levels, that we can turn our thermostats on or off remotely using our smartphones, that our cars warn the dealership when their sensors indicate something needs to be replaced, and that the Amazon website can advise us on products we may like based on our customer profile. In an ideal world, data is a fantastic source of ease, comfort, luxury, and efficiency. However, this ideal world does not exist, and the disadvantages of data applications have become increasingly clear in recent years.

Drawbacks

Looking back, this has always been the case: all new technology offers solutions, but at the same time, it almost always creates problems. One of the clearest examples is nuclear fusion. Its discovery may lead to completely clean energy generation—at least insofar as we are capable of controlling the process. The dark side of this technology, such as total annihilation by nuclear bombs, requires far less control of the process and is, unfortunately, all too familiar. An even simpler example is the first time humans made stone tools. They came with tremendous possibilities, some immensely positive, with chopping wood as just one example—but also gave their owners the ability to inflict far greater injuries on other humans than would have previously been imagined. Technology may not only have adverse side effects, but it is also dependent on the motives of those using it with the potential to be exploited for both good and evil.

There are also two sides to the new information society. Social media offers a formidable tool for people to organize themselves. The bright side of the coin is that individuals can take a collective stand against powerful interests and even organize revolutions. The flip side, however, is that governments and companies also have a formidable tool to monitor people closely via the Internet and identify non-conformist elements. This is a problem not only because it can result in unacceptable forms of control, but also because many governments are not particularly transparent in this respect, to put it mildly. As such, we are seeing that the new information society also has drawbacks that are becoming increasingly visible. There are three elements to these drawbacks, which we discuss below.

Privacy Breaches

Big Brother. The term—a reference to the famous book *1984* by George Orwell, which describes an all-knowing government that monitors its citizens in everything they do—seems unavoidable when discussing the future of the information society. The concept is not even particularly strange in an age in which the media informs us nearly every day that governments and companies are neglecting and invading our privacy. In 2013, Edward Snowden became the undisputed symbol of the fight against an increasingly pervasive surveillance state. His story is well known: as a whistleblower, he publicly disclosed that, since 2008, the US National Security Agency (NSA) had broken privacy rules or otherwise overstepped its authority thousands of times. This mainly concerned the unauthorized bugging and tapping of the data traffic of Americans or foreign nationals in the US. There was mass outrage, including from other world leaders—some of whose phones had themselves been bugged.

The information released about the NSA's activities increased the level of resistance people had to governments' appetites for data and the manner in which this hunger was being satisfied. To many, the revelations seemed to be a confirmation of a vague feeling of unease they had for some time.

These feelings are directed not only towards governments, but also towards 'data-rich' companies such as Google, Samsung, Apple and Facebook. There is a reason why these companies regularly clash with supervisory bodies on how they deal with personal data. One of the criticisms is how these companies combine data, such as in the case of Google, which has access to information from various services, like YouTube, Google+ and Gmail. Combining information garnered from different applications is only allowed in Europe (and some other continents) within the boundaries known as 'purpose limitation'. This means that when personal data is collected for a specific purpose, this data may not be used for any other purpose unless the person in question has given explicit authorization.

Painful incidents involving marketing also fuel peoples' negative attitudes. A case that crops up regularly in discussions on how data is used by companies is the story of an American father who only found out his teenaged daughter was pregnant when the retailer Target sent her coupons for pregnancy products.[1] The supermarket had used its sophisticated algorithms to determine her status based on other purchases. Needless to say, this did not go down well. For retailers, having access to data that can be used to extrapolate such real-world information is extremely valuable in the battle against their competitors, because it allows them to target particular customer wants and needs which, in turn, breeds loyalty. Research shows that during life-changing events such as a pregnancy, people have a tendency to change their behavior and consumption patterns.[2] The first supermarket to recognize that these life-changing events are underway wins the jackpot and may have gained a customer that will remain loyal for many years.

The teenaged daughter had some explaining to do at home. Since then she has—probably reluctantly—become something of a symbol of the possible adverse consequences of companies analyzing (personal) data. The Target example makes it painfully clear that human behavior can be predicted with startling accuracy, much more easily than most people would think. Information about our lives is everywhere, with people performing their own forms of analyses. For example, you can pretty much deduce from status updates on Facebook the odds of a relationship coming to an end.[3] This is no surprise to mathematicians. Stephen Wolfram, a famous mathematician and physicist, who developed among other things a next-generation search engine (Wolfram Alpha), put it bluntly, "People are more predictable than particles." [4]

Stories such as the Target one are often shared at the water cooler where everything is thrown into the same pot, with only one possible conclusion: businesses and governments are evil. They don't care about my privacy; their only interest is in making money out of me (in the case of businesses) or exerting more control over me (for governments). And if businesses only use Big Data commercially in order to increase sales, resistance and resentment will probably only continue to grow.

Lack of Transparency

The resentment people feel is not only because of painful incidents such as the one described above, but is also fed by the lack of transparency on the part of businesses and governments. Because they are not being open about what they are doing with our data, they are not only breaking the law in some cases (whether or not they are aware of it), but they are also stoking resistance. This lack of transparency generates an enormous imbalance in society's information flows: businesses and governments are getting to know us better and better, but we have to continue to guess how they operate. The question is for how long society will be willing to accept this lack of balance.

Here is an example: at the end of 2014, certain smart TVs made by the Korean company LG were collecting information and forwarding it to the company's servers. This included information on which channels the television owner was watching, when the owner changed channels, and which programs were being stored on any media connected to the television by a USB plug. A blogger discovered this by researching the Internet traffic of an LG television.[5] He initiated his research after seeing a commercial aimed at potential advertisers in which LG was promoting the possibilities of targeted advertising based on the user data collected. On the consumer side, LG was much less transparent and, according to the blogger, the television even sent information when the consumer had switched off data sharing. The defense given by the company afterwards was not convincing. The lack of transparency about the data collection—the company had not clearly informed buyers of the television's functionality—resulted in significant criticism of LG. For LG, it resulted in the firmware having to be adapted and a lot of reputational damage.

Strategically, it makes complete sense that television manufacturers should want to know our viewing habits. By using the resulting insights they can, at least in theory, provide us with a better service. But obtaining information secretly is not the answer. When companies are not transparent about collecting data from us and their use of such data, and do not offer their customers the choice of opting out of sharing data, they run the risk that their customers will turn against them. In recent years, that risk has been made evident by a determined group of privacy advocates who are out to name and shame organizations that (seem to) misuse our data.

Making Money from Personal Data

For many companies, data is the new gold. By using it cleverly, they can introduce new products and services, organize their activities more efficiently and offer their customers tailored products. Since personal data is worth so much to companies, can just anyone turn it into gold?

To answer this, we first have to clarify who owns it. In some cases—for example, your own personal data—this is completely clear. In other cases, it is more complex: who, for example, owns the data containing the information that you clicked on a certain link on a certain website at a certain time? Do you own that? Or is it the property of the organization operating the website? To date, such discussions have only taken place in the background.

If we were to assume that companies such as Facebook and Google are using data that is not their property, we could simply ask them to pay us for it. That sounds nice—at least it seems reasonable from the user's perspective! However, don't expect this scenario to become main stream any time soon.

One of the reasons why this will never become reality is that we already receive a (hidden) reward. We are already compensated for making our personal data available and for the tracks we leave behind. For example, we receive discounts from loyalty programs and linked (reward) cards. Or what about free e-mail? When Google announced at the time of the introduction of Gmail that it would provide all users with a gigabyte of free storage, there was much disbelief and many even speculated that it was an April Fools' joke. But it was serious. A decade later, it seems completely normal, and we are barely aware any more that this service is a reward for the data we allow Google to use.

There is a second aspect. If companies were to have to start paying us for (access to) our data, this would be an extremely complex process, with an allocation key being required, because not all data would be of the same value. Although this could be resolved, experience shows that the majority of people do not want more complexity but, on the contrary, want more ease of use.

Attempts are being made to introduce new models, but the current initiatives appear likely to appeal only to small niche markets. One example is the company Cayova (an abbreviation of 'capture your value') which started a social network focused on people selling themselves as advertising targets. The public at large is not (yet) embracing this idea. Another initiative is the startup Datacoup, which pays for access to social media profiles and credit card transactions. The resulting insights are sold to businesses and in return, the user receives $10 every month. It is also possible to earn more if users are willing to give up their privacy almost completely. Luth Research reads its users' computers and smartphones, gaining insight into, for example, their search terms, surfing behavior and social media profiles, and also wants users to answer questions about their purchasing behavior. These users can earn up to $100 a month.[6]

We do not see a great future for the large-scale commodification of turning personal data into cash, because it is just too complex, but also because the vast majority of users basically accept the current situation. Therefore, for now, companies such as Facebook and Twitter have a truly enviable business model. They are making money out of something that is not theirs: our data. We accept this en masse, but paradoxically, it also feeds our negative feelings about how companies deal with Big Data.

The Wrong Approach

Without a doubt, privacy is an extremely important issue with regard to Big Data. Nevertheless, there is something funny about the way we talk about it. We seem to end up with almost completely polarized views, from the proponents of the advantages of Big Data on the one hand and the advocates of privacy on the other. Is this stopping us from getting to the heart of the issue, and finding the right approach to ensure our privacy?

We can illustrate this by using an analogy of an event in a completely different domain. In 2011, Anders Breivik detonated a bomb in a van in the center of Oslo and some hours later he shot and killed dozens of people on the island of Utøya. It emerged later that the bomb was made from fertilizer. The ammonium nitrate in the fertilizer can be released pretty easily and, together with an explosive such as diesel oil and a detonator, it can cause a relatively effective explosion. In the months before the attack, Breivik had bought six tons of fertilizer without attracting any attention or suspicion. With his farm as cover, the purchase of such an amount appeared completely normal.

What can we do with this knowledge after such a horrendous event? Should we ban fertilizer? Introduce a maximum amount that can be purchased in any one transaction? Set up a strict control system to monitor sales of fertilizer? Implement tight security measures around the purchase of fertilizer? Forbid the online publication of instructions on how to make a bomb out of fertilizer?

It is not surprising that these questions have hardly been asked, but it is worth noting that none of these suggestions would have prevented the attack as there is no failsafe way to stop people gaining access to either products or information. We should point out that Congress in the US is demanding tighter legislation in this respect.[7] Moreover, fertilizer is a very common product and imposing severe restrictions around its sale in the hope of preventing it from being used to make a bomb is like using a sledgehammer to crack a nut. We do not want to make the normal everyday use of fertilizer almost impossible because we are afraid that one lunatic will misuse it.

Misuse

However, this is exactly what has been happening in recent years regarding privacy in the context of a society in which Big Data is playing an ever-increasing part. New privacy risks are emerging in a society in which almost everything is becoming measurable. The strange thing is that this is one of the few social domains where (legal) measures are being taken to restrict the 'normal' use of data based on the fear that someone will exploit its potential abnormally. The legislation is focused on limiting the collection of data rather than preventing its improper use. In terms of the terrorist Breivik and his fertilizer, this would be the equivalent of legislating to restrict the purchase of fertilizer but focusing little or not at all on the prevention of bomb building and detonation.

A secondary effect of this peculiar focus is that, in the debate about privacy, we actually hardly ever discuss privacy and instead focus mainly on information security. In itself, there is nothing wrong with focusing on information security—indeed, proper information security is essential in this day and age when everything is connected with everything else—but it is not the core of the challenge we face with regard to privacy. What we should really be discussing is how to ensure that the privacy rights of individuals are respected in a world in which ever more information is being collected about us. Instead, we are limiting ourselves to the question of how this data should be stored and secured.

When our focus in the privacy discussion is entirely on strict conditions for the storage of personal data, we may even be impairing the preconditions required to properly protect our privacy, because we are not seeing the other aspects of the issue. This is because having access to personal data is often not even necessary to breach someone's privacy. Again, an analogy will help clarify this. Say Pete goes to the same bar three times a week; he's probably welcomed as a friend by the proprietor, who probably exactly knows what brand of beer Pete likes to drink and maybe also that he has to be protected from himself at the end of the evening. So far, so good. No one sees any privacy problems in this scenario. This may change, however, the moment Pete takes his children into the same bar for a coffee. Pete may not want the bar owner to show how well he knows him, and he definitely doesn't want the bar owner to reminisce about how Pete fell off his bike last week because he drank too much. This would feel like an enormous invasion of privacy. Note that the bar owner has not stored any of Pete's personal data.

Again, we jump to the world of Big Data. In this world, data is often only valuable because a useful insight can be obtained by combining data from various sources. A chain of furniture stores can pick up and send Wi-Fi and Bluetooth signals from smartphones and, for instance, establish that a phone has been brought into the store for the third time this week and its owner has paused in front of the same couch each time. This could result in a discount being offered without the name of, or any other personal data about, the smartphone's owner being known to the store. The proprietor of the bar can do something similar by giving Pete, as a loyal customer, a beer on the house. No advanced technology is required to do that. However, there is a big difference: hopefully the bar owner has his own set of rules and values about how he deals with his customers—and his customers' privacy. Information processing systems do not have this trait as a built-in feature. What is required is that the designers of these systems—and/or the analysts that work with the data—are provided with ethical guidelines on the use of data; and that use must be supervised.

In the above examples, we discussed the collection of (large volumes of) data, which are known as 'implicit identifiers' (for example, signals picked up from smartphones). It appears that discussions about this kind of data are rarely focused on how it is used, but almost always about the possibility of turning this anonymous data back into personal data, such as through the use of data matching or similar techniques. This de-anonymization process is known as re-identification. But what exactly is personal data? When is a point of data traceable to a person? The challenge in this respect is that a single point of data cannot be traced back, but has to be combined with other data. Whether or not a point of data is traceable depends on the circumstances. And let's not even get into a discussion about what exactly traceable means. Because what if the bar owner doesn't know Pete's name and address. Does this mean his data is untraceable?

Often, these nuances are neglected, with many privacy advocates focusing on limiting the storage of anything that even smells like personal data. Although their intentions are good—fighting for the rights of the individual—this approach cannot hold up in a world that is increasingly data-driven. It could even be damaging, for example, by hampering large-scale research into various pathologies.

The reality is that organizations are often not too interested in personal data and, in many cases, only want aggregated and statistical data. To continue the analogy with the bar owner: he doesn't need to know Pete's last name or where he lives, as long as Pete feels at home in the bar and returns as often as possible. Although it is not often acknowledged that the discussion on privacy is too narrow, the fear of the unknown is accompanied by a reflexive urge to return to existing patterns, as demonstrated by a petition signed by one hundred European scientists, which states:

"Technically, it is easy to relate data collected over a long period of time to a unique individual. Economically, it may be true that the identification of individuals is not currently an industry priority. However, the potential for this re-identification is appealing and can therefore not be excluded from happening."[8]

Prohibition

Fear can sometimes override more analytical thinking, which brings us back to the fertilizer. Based on this reasoning, limiting the collection of data translates as prohibiting the purchase of fertilizer because people such as Breivik exist. This just doesn't make sense. If we want to create a world in which we can capitalize on the massive potential advantages of Big Data in a controlled manner, we cannot try to prevent all potential future abuse by prohibiting normal everyday use.

Only when we focus on creating insights from personal data in a controllable manner, rather than (only) worrying about storage, can we make real advances towards protecting peoples' privacy. Only then can we have a discussion about the ethics of the usages we agree or disagree with and organize proper supervision of the analysis and application, not (only) the storage, of personal data. We discuss this ethical dimension further in Chap. 8.

Conflicting Behavior

Earlier in this chapter, we included some examples of how companies' and institutions' appetite for data has resulted in a number of unwanted incidents. It isn't hard to imagine that a growing number of businesses and governments will face such incidents in the coming years. It therefore seems likely that public resentment will increase, rather than decrease. Such incidents have mainly occurred in the 'new economy' of Internet, computer, and electronics companies. It is these companies that have often thought longest and hardest about this theme and push the limits.

Many companies that matured in the 'old economy'—banks, insurers, energy companies, etc.—have hardly even begun to think about whether they are handling data properly and whether they are sufficiently transparent for their customers. Only now are they starting to connect data 'silos' and realizing (although they are probably not realizing the extent of the issue) that, in doing so, they may be taking steps that will face public resistance.

A good example of this was the explosion of outrage[9] that followed the announcement by the Dutch bank ING of its intention to make customers personalized offers based on their transaction histories. It was only a pilot project, but all of a sudden, all 17 million Dutch people had an opinion on how ING was handling their personal data, and a pretty strong one at that. The consensus in populist discussions, such as those on Twitter, was that the bank needed to stop the project. Immediately. A number of national politicians even threatened to close their ING accounts. A few days later, the bank succumbed to public pressure and withdrew the plan (for now).

Banks are probably particularly sensitive to this kind of incident since the financial crisis—it is not for nothing that, in 2013, five years after the start of the credit crisis, banks and financial service providers were still the least trusted sector.[10] Even more damning was a survey of 10,000 millennials that resulted in a list of the top 10 most hated brands. Four of those brands were banks.[11] Other sectors are also facing similar criticism. In the UK, hospitals sold medical data to insurance companies on a large scale, understandably causing a public outcry.[12]

There is a big difference in this regard between the US and Europe. Americans are more accustomed to the fact that companies use their data for various purposes, and the use of data is therefore a much less sensitive issue. Nevertheless, even in the US, companies may face significant reputational damage if their zeal for data collection goes too far. A good example is the protest that occurred when it became clear that OnStar, a subsidiary of General Motors, which collects GPS data from cars—including on behalf of insurers—had changed its conditions. According to the new 'terms and conditions'—which are often routinely accepted without being read by users—data collection would continue even after the user's account had been terminated. When this became common public knowledge, there was an outcry, with politicians also getting involved. Senator Chuck Schumer said it was "one of the most brazen invasions of privacy in recent memory."[13]

This is only a snapshot of the many indications out there of significant resistance to the new forms of data consumption. It is not surprising, therefore, that discussions on talk shows, in bars and at the coffee machine at work about such (privacy) incidents often become heated. How is it then possible that companies and governments who have so grossly abused our data have gotten away with it? Edward Snowden was (and is) a hero to many and has opened our eyes. The gist of most peoples' reactions to his revelations has been that this has to stop.

However, something strange is often going on with these discussions. Five minutes later, we are back to talking about football or that funny viral on YouTube and checking our social media as if nothing happened. We may even post a selfie, or tag our friends so that they know which café we're at. We may even post how many beers we've had...

The moral of the story is that Snowden has (so far) changed very little about our behavior. We are angry, but we can file that anger away as a memory very easily. When an even cooler app, smartphone or smartwatch comes along, we'll probably want that too. We get worked up for a second and then it's back to business. How can we explain this conflicting behavior? Let's list some factors.

Everyone Loves a Freebie

"There's no such thing as a free lunch." These are the oft-quoted words of the American economist and Nobel Prize winner Milton Friedman. Many products and services may seem free, but they never are—there are always hidden costs. The 'free' economy has always existed but, with the rise of the Internet, it has become much more prevalent and is contributing to our contradictory behavior.

"$0.00 is the future of business," said Chris Anderson as far back as 2008 in Wired.[14] He differentiated six business models that provide free products or services. One of those, the advertising model, has become dominant on the Internet over the last

10 years. The content, software, or service is free, but the user is exposed to the advertisements.

The advertising model fits the nature of the Internet perfectly. First of all, the number of people who see a certain ad and the click on it can be measured precisely. Advertisers therefore no longer have to pay for ads or commercials that no one looks at. But a second aspect is at least as important: the Internet makes tailored advertising possible. Because Internet companies can now learn about your likes and dislikes, they can tailor the adverts to your taste. The advertising model—invented by newspapers and magazines—has therefore been refined on the Internet over the last 10 years. The late Freddy Heineken, former chairman of the board of directors and CEO of the brewery Heineken International, allegedly used to complain that half his advertising budget was wasted but that the real problem was that he never knew which half. The Internet has at least eased that pain somewhat.[15]

For us as consumers, it's also good news. We don't have to hand over any money to use the best search engines, music services or handy apps. We can pay to use these services with our data. However, we need to realize one thing: if something is free, most likely we are the product.

In this regard, it is intriguing to look at the belief held by Google founders Larry Page and Sergey Brin when they were studying at Stanford University and how this has evolved over the years. While at college, they wrote, "We expect that advertising-funded search engines will be inherently biased towards the advertisers and away from the needs of the consumers."[16] Compare this to the 2014 Google philosophy: "Focus on the user and all else will follow."[17] This is in huge contrast to their vision as students, when they did not think it would be possible for the user to be central in the advertising model.

We Are Creatures of Habit that Follow the Herd

Humans are not always rational beings making well-considered choices. Often, we're creatures of habit, stuck in routines that are nearly impossible to change. This is part of the reason why people find it so difficult to break bad habits such as smoking, eating junk food and not exercising enough. We know that we have to change and we will change. But not today. Tomorrow.

Routines are essential to our lives due to the limits of our brain capacity and, because we are basically creatures of habit, we are comfortable following these routines. Scientists claim that our brains are continuously searching for ways to lighten the load and our habitual behavior is very helpful in this respect, because habits don't require much brainpower. The book *The Power of Habit: Why We Do What We Do in Life and Business* by Charles Duhigg describes how a woman turns her life around completely. She quits smoking, changes jobs and starts running marathons. Scans show that the patterns in her brain have also fundamentally changed. But basically, her brain has exchanged one autopilot setting for another.

We are not only creatures of habit, but also pack animals. This is the cause of a significant amount of both fraud and questionable business practices. The reasoning, whether conscious or not, is as follows: if everyone else is crossing a line to get a customer on board, it's okay for me to do it too. This was the case before the last economic crisis, when almost everyone in the financial world seemed to think there were hardly any risks involved in financial transactions anymore; and it was almost impossible to disagree as an individual. The habitual behavior in our immediate environment—that of colleagues, customers, etc.—greatly influences our own norms and values. People mostly learn their norms from the behavior of people they see as role models. This is aligned with the theory of cultural relativism,[18] which tells us that there is no universal truth on which to base our ethics, but that our interpretations are significantly influenced by culture and therefore our environment. We have almost all been the victims of herd behavior at one time or another. The Dutch brain research scientist, Victor Lamme and the late Harvard professor of psychology Daniel M. Wegner, even go so far as to say that free will does not exist.[19] According to them, we mainly imitate other people and follow the instincts of our brain, while fooling ourselves into thinking that there are explanations for our behavior. In reality, however, according to Lamme, our thoughts follow our actions and not vice versa. One of the examples he uses is the cards you find in hotel bathrooms asking you to help save the environment by leaving only dirty towels on the floor, and towels that are still clean on the hook. The more clearly the cards show how many guests respond positively to these requests, the greater the willingness of others to cooperate.

We Are no Better Than the NSA

According to some, we are already living out the classic doomsday scenario of a surveillance state, with a multitude of surveillance cameras on the streets and disconcerting government tapping practices. However, the reality is that we are completely complicit in this. George Orwell's *1984* Big Brother scenario assumed an all-powerful government that had eyes and ears everywhere. The reality is that we are the biggest Big Brother. We continuously record our daily activities—and those of our friends—with our phones; we upload photos to Instagram or Flickr and videos to YouTube and Vine. On-board cameras in cars mean we can even 'enjoy' the accidents of our fellow road users (the latter being very popular in Russia in particular).

The real issue is not so much surveillance, but sousveillance. Surveillance is about an entity watching us from above ('sur' means 'from above' in French), while in sousveillance, the users transmit information upwards from below ('sous' means 'under' in French). It became clear that sousveillance is a formidable tool during the search for the perpetrators of the 2013 Boston marathon bombings. A collaboration between users of the social networking site Reddit very rapidly resulted in a manhunt for a missing student who had, completely erroneously, been identified as the perpetrator of the attack.[20] These amateur detectives were not equipped to assess whether the 'suspect' was the right person and did not think what the consequences of their actions might be. Without any real evidence, their judgment had already been made. There is only one possible conclusion: we are turning ourselves into Big Brother.

We Are Addicted to Personalized Information

We expect Google to provide us with relevant and useful information that exactly matches our personal needs. In order to do this, Google needs to know more about us. The more it knows, the more relevant the information it can give us. From a music service such as Spotify, we expect our favorite playlist to be available on all our devices and maybe even to receive advance notice of a concert nearby that we may like to attend. For this to be possible, we have to share data with Spotify.

Personalized, tailored services seem to be the norm for every service provider on the Internet. We have become addicted, and our expectations get higher all the time. Ideally, we would like our search engine to know what we want before we do, our mail program to prevent embarrassing mistakes and our music service to have already ordered the tickets for a cool concert before anyone else even knows that our favorite artist is coming to town. Easy as pie. But these desires can only be fulfilled if we are willing to give all these companies access to our data.

Tension Between the Pros and Cons of Big Data

There is clearly tension between what we see as the pros and cons of Big Data. On the one hand, Big Data offers big advantages and we, as users, have massive expectations about how organizations should make those advantages available to us. On the other hand, there is a lot of resentment about the accompanying downsides and even the whiff of Big Data tends to meet resistance. It is human nature to resist change and this resistance is fed by the scope (and speed) of the changes—which are huge in this case and definitely disruptive. How can businesses and governments ensure they end up on the right side in this tug of war?

Offering Social Value

Remarkably, many (big) companies are still not making real choices in respect of the fundamental changes that Big Data is bringing their way and how they want to deal with it. Not choosing at all is the worst possible choice. They are holding onto what they know and are afraid of losing existing business should they decide on a (radical) change of course, into the unknown.

History has demonstrated many times what the risks are when companies are incapable of embracing new and disruptive concepts. Kodak missed out on the rise of digital photography, even though the company had the knowledge it needed to become successful in this area. The smartphone brought Nokia to its knees. WhatsApp made texting practically obsolete. Music streaming services turned the music industry upside down—or at least that part of it that was unable to keep up with a changing digital music landscape. It's possible that Bitcoin will do the same to banks. Bill Gates is quoted as saying: "We need banking but we don't need banks anymore."[21] but more on this subject in Chap. 6. And what if self-driving cars actually became a fact of life for ordinary road users, not just a Google research project. This would have an enormous impact on both business and industry. In such a world, would car insurance even be required?

These examples are closely connected to the new possibilities afforded by Big Data. The question remains, however: how can companies embrace Big Data without meeting resistance from the public and thereby shooting themselves in the foot?

In essence, the answer to that question is simple. At the start of a Big Data initiative, every organization needs to consider the value to the user. If only commercial gain is considered—how can we sell more diapers, advertisements, cars—or if the benefit of using the data becomes difficult to justify, sooner or later companies will end up having to deal with a serious incident. Therefore, companies need to be sure they are not only going to create financial value—make profits—but are also going to create social value, offering the user an additional benefit. Such companies will then have a future-proof strategy, tapping directly into the zeitgeist.

The prominent Harvard professor, Michael E. Porter, outlined this approach in an article in the *Harvard Business Review*[22] with the title *Creating Shared Value*. In this concept, a company pursues not only economic profit but also value creation for both people and the planet. According to Porter, future businesses will adopt models under which they create social value that may also result in financial profit.

This is not a naive assumption, but a trend that is already visible from the early adopters. So far, the leading examples of such business models have had little to do with data. For example, Unilever is focusing on reducing child mortality due to poor hygiene in poverty-stricken areas in Asia[23] by providing special soap products and education. This gives the company a moral license to operate because, in addition to commercial value, it is also creating social value (and at the same time, demand for soap is increasing). The reaction of Apple CEO, Tim Cook, during a shareholders' meeting in March 2014 was a clear case of connection with the new zeitgeist. A conservative investor criticized Apple's decision to invest in green energy projects, because they do not result in immediate financial profits. Cook was infuriated by such shortsightedness and had some advice for the investor: "If you want me to do things only for ROI (return on investment) reasons, you should get out of this stock."[24] Cook made it absolutely clear that he is targeting both social and financial profit.

Offering Social Value in Practice

This same combination should also be key when using data, but how does it work in practice? Here are some examples:

MasterCard wants to increase credit card security by only allowing payments when your smartphone is near your credit card.[25] The company is performing pilot studies using the geo-data of mobile phones (which, in many other cases, is leading to a lot of comment about privacy issues). Commercial profit: lower fraud costs. Social profit: safer payments.

Snapshot is a device provided by the American insurer Progressive that can be plugged into a car's diagnostic port (usually on the underside of the steering column). The device monitors driving behavior—speed, braking behavior, distance travelled and driving after dark—and the Snapshot data is then transmitted to the insurer. The advantage? The 'better' the driver, the lower the insurance premium. It seems a very smart way to (financially) motivate young drivers to drive more carefully. Commercial profit: a lower premium. Social profit: improved road safety.

If we are willing to share our lifestyle data with our health insurer, this may contribute to the early detection of health problems, or to scientific studies aimed at finding treatments and cures. Commercial profit: lower health care costs. Social profit: better health. We discuss this in more detail in Chap. 4.

A Public Backlash May Never Be Far Away

We acknowledge that it may not always be easy to transfer the combination of financial and social value from the drawing board to daily life. An incident involving the Dutch satellite navigation company TomTom provides a striking example. This company owns a wealth of (actual) information on the driving behavior of road users. Its users are happy because, based on their selected destination, TomTom immediately suggests an alternative route if it detects delays from the GPS measurements it is collecting in real time from thousands of TomTom users. Local authorities also use this information: it gives them clear insights into traffic bottlenecks and how to resolve them. This also benefits the driver.

In 2011 however, it emerged that TomTom was selling the aggregated data it collected about car speeds, via a specialist government agency to the police who said that it helped them to obtain insight into the most dangerous traffic situations. When it emerged that the information was also being used to determine where to place speed checks, there was nationwide uproar.[26] People were furious that TomTom had allowed this to happen, even though the data was anonymous and aggregated and no private information was involved. TomTom might believe it was providing a social benefit by contributing to greater road safety. However, the public thought otherwise. TomTom immediately stopped providing the data, apologized and changed its policy.

The commotion surrounding the use of payment data by ING mentioned earlier also shows how careful companies have to be. The bank received a tsunami of negative criticism after a test was announced that would enable the bank to offer personalized promotions based on customers' spending habits. ING is a symbol for the struggle that many other banks—in both Europe and the US—are going through. We should also point out, however, that a small number of opinion columns and weblogs internationally praised ING, because the test was exploring ways the bank could create actual added value for its users. ING's clumsy and confused communication about its plans probably played a significant role in the resultant national outcry.

During a second attempt to explain its plans to the public at large, the bank emphasized how it was trying to understand customer behavior better, based on payment data, with the potential advantage of improved customer service. Was this second attempt a real clarification of its plans or more of a rebranding exercise? There's no way to know for sure. According to ING, it was never its intention to sell customer data to third parties— which was the issue that met so much resistance and continued to provoke so much online discussion long after the bank had backed down. The bank also announced that the plan would have been discussed extensively with customers and other stakeholders before any decision was made about its implementation.

The entire affair was a major public relations disaster for ING. It has left its customers with a bad aftertaste that will not fade overnight, even though ING claimed to be acting solely in customers' best interests. This case shows how vital it is for organizations to clearly communicate the intention behind their Big Data projects.

A Devilish Dilemma

It is clear that the wealth of insights into our actions that is available to businesses is becoming an increasingly important aspect in competition. Eric Schmidt, the former CEO of Google, put it strikingly in an interview in 2010 with *Der Spiegel* in Germany: he sees a future in which machines and technology play an ever-increasing role and he wants Google to know as much as possible about us, simply in order to improve our search results. "We won't need you to type at all. We know where you are. We know where you've been. We can more or less know what you're thinking about. [...] Google policy is to get right up to the creepy line ... but not cross it".[27] Many Big Data plans face a major dilemma. Companies need more and more data in order to create (social) value to provide the services we require. The more we are willing to share that data, the more they are capable of meeting our needs and thereby creating social value. But many people are strongly against sharing more data. They associate Big Data with Big Brother scenarios, and are worried that companies only want to make as much money as possible from our personal data and that governments don't care about our privacy. We can't blame them, as these feelings have been fed in recent years by numerous scandals. Time and time again, it has been shown that it is essential for businesses and governments to make it clear to the public why they are sharing their data, so they can decide for themselves if this is a good thing. Those who can't do this are unlikely to be successful in the long run and may as well shelve their plans before they even start.

Chapter 3
You Ain't Seen Nothing Yet

How everything will be turned upside down in the new information society

Send Your Kid to Dental School

There's a 94 % chance that the accountancy profession will die out in the next 20 years simply because accountants' jobs will be taken over by computers. Nor are you safe as a real estate agent, with an 86 % likelihood of extinction. A better choice is fire fighter, with a 17 % risk. Or even better, dentist, at 0.004 %.[1]

In the first chapter, we showed how the rise of Big Data is affecting almost every aspect of society. From how we get around, to how we grow crops and everything in between, all our wishes and needs are being met by tailor-made solutions. However, at the moment these changes are still relatively small adjustments to existing concepts. The question is, therefore, whether the new information society will really be all that different from the society we're living in right now.

In our opinion, the answer is a resounding yes. The new information society will be a radically different world in many ways, which we discuss in this chapter and the next. In this chapter, we present two examples: the rise of virtual currencies such as Bitcoin, and the impact of new technology on our jobs. These two areas of change show how invasive the consequences of Big Data can be. The next chapter, Chap. 4, describes the potential for change in health care. The changes in this sector are so fundamental—and meaningful to society—that they justify taking up a whole chapter.

© Atlantis Press and the author(s) 2016
S. Klous and N. Wielaard, *We are Big Data*,
DOI 10.2991/978-94-6239-183-3_3

The Rise of Bitcoin

As far back as 1999, economist Milton Friedman predicted a world in which banks would no longer play a part in payment traffic.[2] And, when you consider the rise of Bitcoin and other cryptocurrencies, he might be right. Bitcoin is a digital currency that allows people to transfer money from their electronic wallets into the electronic wallets of other people without any intermediary. 2013 seems to have been the year in which Bitcoin—at least in terms of publicity—made a real breakthrough. There was a lot of discussion about the potential of the concept. Alan Greenspan, former president of the American Federal Reserve, could not see any value in Bitcoin.[3] Paul Krugman, a former Nobel Prize winner and prominent columnist in *The New York Times*, was also having none of it.[4] The volatility in the currency's value and various scandals and incidents involving major players in the Bitcoin ecosystem (Mt. Gox and Silk Road) would appear to back them up in these opinions. So it seems that a large group of prominent thinkers has more or less relegated Bitcoin to the scrap heap.

At the same time, there are numerous people and bodies embracing the currency. The entrepreneurs, private individuals, and organizations who were early adopters say it has a lot of potential. Are they stupid or can they see something the intellectuals can't?

Perhaps both sides are right. In order to get a clear picture we need to break it into its component parts: into the (retained) value of Bitcoin as a currency, and the Bitcoin mechanism as a different way of doing business.

To start with the first point, Bitcoin—or a comparable cryptocurrency—is, in fact, not a currency but a raw material. The big difference is that currencies can be printed endlessly by central banks while raw materials such as gold have a finite stock. At a certain point, they are just gone. This is also the case for Bitcoin, although this raw material is mined completely electronically with a computer. Bitcoin's algorithm determines that there are a maximum of 21 million Bitcoins and as time progresses, they get harder to mine—which means that finding an undiscovered Bitcoin requires ever more calculation power. When we look at Bitcoin like this—as a raw material such as gold—we have to accept that its value may fluctuate significantly. We cannot rule out that its value may increase or decrease significantly in the future, depending on demand. Only time will tell: any predictions are just guesswork.

However, the mechanism behind Bitcoin is a totally different ballgame. The key question here is not what the value of Bitcoin is, but what the world will look like when individuals can do transactions with each other without any need for a bank or a notary. Because that is the essence of the Bitcoin mechanism.

In order to start understanding it, we have to take a peek under Bitcoin's hood. Satoshi Nakamoto, a Japanese man so mysterious we aren't sure whether he is actually even a person, was widely acknowledged as Bitcoin's designer. He devised an extremely clever mathematical solution where the certainty of a transaction is derived from the system itself and not from the (intermediate) parties involved. In its simplest form, this is possible because every user participates in checking the transaction. Every user has his or her own overview, on his or her own computer or smartphone, of all the transactions between all the other users. This database of transactions is known as the blockchain and all these overviews collectively constitute the control mechanism for each transaction,[5] providing a shared ledger. In other words, it is possible to do reliable business in a system where there are unreliable partners!

The real revolution is the blockchain, the mechanism behind Bitcoin that makes transactions without a bank or notary as a trusted intermediary possible. This is because the network itself constitutes the trusted party that supervises whether the agreed transaction has indeed taken place in line with the agreed conditions.

Is Bitcoin to Banks What Torrents Were to the Music Industry?

There is an interesting analogy between Bitcoin and torrents, the extremely popular way to (illegally) download music and movies. In the case of torrents, no central computer or middleman is required either: when you download a music file with a torrent, it comes in pieces from different computers belonging to other users. And yet, the file is downloaded as a whole. The decentralized network is obviously capable of doing that. These networks, which are known as peer-to-peer networks, have been stable for many years and the music industry has a hard time fighting them because of their decentralized character. The Internet community truly came up with a powerful new way to share music. Meanwhile, the music industry spent too long in denial, initiating numerous court cases to try to stop illicit downloading and getting into financial difficulties in the process. In recent years, the movie industry has faced a similar challenge, with Popcorn Time's software—based on torrents—being the standout package to date.

Here we encounter one of the greatest challenges in this era of huge technological change: the question of what the position of the established order should be. At what point do you decide to embrace a fundamentally new development? Will Bitcoin turn the financial sector upside-down, in the same way as torrents affected the music industry? Will banks be as slow to adapt to change as the music industry was? It seems possible that Internet companies will embrace this concept since there is speculation[6] that Google is developing its own digital currency. Making any firm prediction is difficult. For example, Santander, the Spanish Banking Group, not only recognizes that the blockchain could make bank payment traffic significantly less expensive—its own research mentions an efficiency gain of $20 billion—but also that there is a world of other opportunities. There is a dilemma however, as revealed during an interview with two Santander stakeholders: "This will only be interesting if numerous banks take part and collaborate. We are talking and experimenting with several banks" and "It's like having the first phone—there's no point, you can't ring anyone."[7] Blockchain is also receiving attention in the world of trading. NASDAQ, the business behind the NASDAQ exchange and the owner of various European exchanges, recently started experimenting with it.[8]

In any case, there is no doubt that the blockchain concept is enormously powerful, because the system can take over the role of an independent third party. Any transaction requiring a signature can be handled by this network, from share transactions to marriages and divorces. Traditional financial sector organizations should start to think about what their added value may be once new concepts of this type break through. Teething problems, such as the instability of the network, developments in the protocol and unreliability of parties in the network will not change this: the genie is out of the bottle and there is no getting it back in.

The genie is becoming even more powerful since the blockchain is becoming visible in an increasing number of areas. An example is Storj, an initiative what can be best described as a cross between Bitcoin and Dropbox. The team behind Storj is developing a storage service in the cloud without central servers but which stores the files of its users in a decentralized manner in small sections on the computers of the users themselves, as with the popular torrent model. The big difference is that, in the case of Storj, the storage is secured by applying the blockchain. Another example comes from Vitalik Buterin, a then 19-year-old American hacker. He sees the blockchain not only as a secure and reliable method for financial transactions, but has also used it to build his brainchild Ethereum,[9] which provides a basis for the decentralized running of almost all digital applications. One of Ethereum's interesting aspects is that it results in a totally different legal reality. Who owns an application that is decentralized? In which country is the company that exploits these applications located and does this company pay taxes? In a certain sense, it creates an alternate reality—an ultimate version of cloud computing—which has no rules defined by borders. It creates a domain that cannot be reached by policy-makers, legislators, or supervisory bodies. It's the closest we are to what we used to call cyberspace.

The End of the Barber, the Real Estate Agent, the Bricklayer, the …

Bitcoin is an example of the truly fundamental impact that Big Data-related developments can have on society. In a very different area, something else is happening which may have at least as much of an impact: the way in which new technology combined with Big Data will make human work increasingly redundant.

Let's go back a couple of years. In 2010, the American economy was recovering after the financial crisis. But there was something remarkable going on: the growth of the economy wasn't being reflected in growth in employment. Famous economists discussed the how and why of this jobless recovery.

This is not a new phenomenon. It had previously been seen in the 1930s. Throughout the intervening decades, as machines were increasingly taking over manufacturing from workers, there has been recurring concern that the technological innovations would result in significant job losses that could no longer be off-set by the creation of new jobs in alternative sectors.

The extent to which IT is so closely interwoven with our daily lives has amplified these worries. For decades now, people have been speculating that computers will take over large parts of our work. This has not happened to date, and the persistent 'productivity paradox' is important in this respect: despite large investments in IT, general productivity has not increased. Nobel Prize winner Rob Solow mentioned this phenomenom rather dryly: "You can see the computer age everywhere these days except in the productivity statistics."[10] Data shows again and again that employment in the manufacturing sector—where IT has an enormous impact on productivity—is falling, but at the same time, there has been an enormous increase in jobs in the service sector—where IT does not seem to make us more efficient. It might not be surprising that research conducted in 2012 by Twente University in the Netherlands showed that, on average, we lose a whopping 8 % of our working time to poorly-functioning IT and poor digital skills. According to the researchers, this results in the loss of €19 billion every year.[11] It should be noted that this is the figure for the Netherlands alone, an economy with a total size of €600 billion. Globally, the lost productivity could amount to hundreds of billions or even trillions of euros.

However, this could all change: information technology may yet lead to radical productivity gains and the associated disruptive effects on society. The magazine *The Economist* already tentatively suggested that the productivity paradox might slowly be becoming a thing of the past.[12] Businesses clearly needed some time to discover how IT can contribute to efficiency. The end of the paradox is doubtless linked to the fact that IT is no longer used only to make the things we do more efficient, but also to set up completely different processes. Without human intervention if necessary.

A much-discussed research paper in this regard is the one published by Carl Benedikt Frey and Michael Osborne in 2013.[13] Using the model they had developed, Frey and Osborne calculated which professions could be under threat now that computers are no longer limited to dealing with routine tasks, but are also able to handle patterns in data and take over more intelligent tasks. They predicted that almost half of all professions could disappear in the next 20 years, including telemarketers, referees, taxi drivers, real estate agents, barbers, and bartenders. And what jobs do have a future? Mathematicians. Oh, and also babysitters, relationship counselors, and recreational therapists.

One of the underlying causes of this is the rise of robotics. This field has been promising earth-shattering developments for years, but is still at a very early stage in relative terms. However, this could change in the next few years, partly because afford-able robots are now being marketed that not only execute programmed tasks reliably, but are also capable of human inter-action.[14] An example is Baxter,[15] a robot that even tries to make eye contact and can be trained intuitively by colleagues on the floor simply by taking him by the 'hand'.

Actually, we're pretty certain that Frey and Osborne's predic-tions will come true. Someday. We know that those jobs will disappear. We can draw a parallel with developments in the agricultural sector over a century ago. Once, agriculture was the source of almost all our work, but today it accounts for only about 2 % of jobs. The rise of intelligent robots is having a similar effect on labor requirements in general.

Redistribution of Wealth

The rise of robotics—which cannot be viewed separately from the 'datafication' of society discussed in this book—means that fun-damental questions will have to be answered. The issue in this respect is no longer that technology allows us to do things faster; the issue is that it simply takes over the tasks from us completely, resulting in a radical productivity gain and a society with almost full unemployment. An intriguing example is the announce-ment by the crowdfunding platform Kickstarter that it would no longer have a person assessing all its projects. In February 2015, the company published data in this respect. Of the 69,000 appli-cations submitted since the new approval system was intro-duced, approximately one-third had been placed on the platform through the 'fast track' without any human assessment.[16] Those who submit a project have to accept that algorithms that evalu-ate thousands of data points will determine whether their project is ready to be launched on the site.

What are the future effects of robotization? It is possible that much of the work outsourced to low-cost countries in the last decade will be repatriated by Western countries because, cost-wise, it makes sense to robotize manufacturing near the product market, saving transport costs and potentially also gaining in terms of quality. This could result in significant unemployment and an increase in poverty in these developing economies. The same applies to the Western world, where most blue-collar jobs are in danger of disappearing. Then, the big question of the redistribution of wealth will have to be answered. And reducing—or at least maintaining—the current unemployment level will only be possible through the redistribution or redefinition of labor. Working times could be drastically shortened. Could we get by with a one-day working week? Or should we buck the trend of recent years and drastically reduce retirement ages?

It seems inevitable that there will be less work in the future. The question is whether this is a problem. In the end, it's about whether we can meet our needs as a society. If that proves possible without us having to work, then there shouldn't be a problem. In any case, this is an issue that requires the attention of policymakers and politicians (and lately it seems to have been finally getting some).

An analysis in the Financial Times shows that at various times in the past—for example, at the beginning of the industrial revolution—we faced similar questions, and also that it is not as easy as it may seem to react to changes quickly. That's because we only really see change when it is already occurring. "If something radically new is happening or is about to happen, there's a chance we won't know for sure until well after the process has started. The societal and political responses might thus come too slowly."[17] Perhaps these changes are already more visible than many realize. Historical data on unemployment rates during and after the economic crisis speak for themselves. Decades ago, unemployment numbers recovered extremely fast after a crisis, often within a few years. Modern-day crises follow a totally different pattern: after a crisis, employment returns to its pre-crisis level incredibly slowly.[18] Professor James Bessen wrote in an article for the International Monetary Fund (IMF) that politicians could anticipate this by developing policies that focus on developing knowledge and competencies for new jobs. He also mentioned that the past offers valuable lessons in this respect: "In the past, training institutions and labor markets sometimes took a long time to adapt to major new technologies. For example, during the industrial revolution, factory wages were stagnant for decades until technical skills and training were standardized; when that happened, factory wages rose sharply."[19]

The speed at which a cocktail of developments—robotization, the rise of data analysis, the development of artificial intelligence—is occurring this time is creating a lot of urgency around the need to develop new policies.

Chapter 4
Hitting the Bullseye First Time Around

Tell me what tumor you have and I'll give you the right medicine

A Square Centimeter of the Mona Lisa

You may have been wondering since the beginning of this book what that blurry picture on the first page is. We have tested your patience long enough. It is a piece of the Mona Lisa, Leonardo da Vinci's most famous work. That picture is there for a reason of course. We wanted to make a statement.

Suppose, you are a philistine and you sneak your way into the Louvre in Paris. Using a box cutter, you cut a square centimeter from the Mona Lisa. You take this little piece of canvas to an art expert and ask him what painting it is. It would clearly be close to impossible for him to say because the art expert cannot see the whole picture and can only determine, based on the materials used, that it is probably a 16th century painting.

Asking such a question is obviously silly, so why is it so very common in health care? Right now, a person with a chronic disease has, on average, five hours a year of personal contact with a specialist. Since an average person is awake approximately 5600 h a year, the specialist is trying to make observations, draw conclusions and find the best possible treatment for the person within one-thousandth of that time. The specialist, therefore, only has a very limited data set on which to base the diagnosis. It's a miracle that it works as often as it does.

© Atlantis Press and the author(s) 2016
S. Klous and N. Wielaard, *We are Big Data*,
DOI 10.2991/978-94-6239-183-3_4

We have shown previously that when we embrace the new possibilities that (data) technology offers, amazing opportunities arise. In this chapter, we discuss what that same attitude can bring to the world of health care. We did not choose this sector without reason: we believe that the potential for data application is extremely high in this particular field.

An Enormous Contrast

When viewed from afar, there is a huge paradox in health care. There is a lot of high-quality scientific research; but we market that high-quality research in a hopelessly old-fashioned manner. This approach is patently obvious in something that touches each and every one of us at some point: taking medication.

Almost all generic medicines are currently based on research into their effect on a (usually young) white Western male.[1] This happens even though it has long been accepted in medical science that women respond differently to medication and also that other characteristics—such as age, weight, and ethnicity—significantly influence the efficacy and the (undesired) side effects of a drug. When you think about, it is sheer madness. Compare this to the example of The Climate Corporation, which we mentioned earlier in this book: clearly, we are capable of providing farmers with accurate advice, based on the composition of their soil and predictive weather models, on how to optimize their crops, but we still use a scattergun approach where our health is concerned. There should be other ways and in this chapter we will investigate them. We will subsequently address:

A radical tailored approach to medical treatments and diagnoses. It's the holy grail in health care: to provide drugs and medical advice that are completely tailored to your personal lifestyle and DNA. Tailor-made approaches are getting closer, but require a totally different setup for our health care systems;

From health care to self-care. By using modern technology (e.g. sensors), we can obtain more insight into our movement patterns, diets, and our bodies' parameters. This will put us much more in the driver's seat of health care. We will be able to actively improve our health and we can continue to provide care to the elderly and people with chronic illnesses properly and affordably;

From care to prevention. Access to proper medical care is not enough. We can build a world in which we focus on preventing health problems. For this to become a reality, the entire sector— from health insurers to general practitioners to pharmaceutical companies—has to get involved.

Radical Tailored Approach to Medical Treatments

Back in the late 1990s, the days of one particular Finnish woman with metastatic colon cancer seemed to be numbered. Until her husband read that a Japanese scientist had discovered an abnormal receptor—a type of protein—on her type of tumor was the culprit. Through his connections, the husband was able to acquire a drug that blocked that particular receptor successfully and, after only a month, the cancer cells were gone. The case was written up in an article in the prestigious *New England Journal of Medicine*, and scientists all over the world enthusiastically embarked on this area of research.[2]

Does this charming story mean that we're on the way to finding the ultimate weapon against cancer? Perhaps; perhaps not. What is more significant is that it is a great example of taking a radical tailor-made approach in medical treatments. The magazine in which this story appeared called it 'cancer's weak spot' and the Dutch oncologist Marc Peeters used the following striking analogy: "When administering regular chemotherapy we use a giant hammer to hit as many cancer cells as possible. With focused medication, we're targeting specific core points within the cancer cell. Unfortunately, these medicines don't work for every patient. And you need to have a clue as to where to start the treatment. So for each tumor the first step is to find these clues."

In short: a hammer—the chemotherapy—is not the best tool for all types of cancer. The example symbolizes the challenge in a wider sense: we have to obtain a much better insight into our bodies. Because the more a medical specialist knows about us the better the treatment can be. The combination of clinical data with data on our diet, our lifestyle, etc. offers the key to a future with optimum medical care. New technologies will enable us to continuously monitor our health using various sensors (in, on or close to the body). This will make available data that provides specialists with valuable insights which allows them to develop unprecedented personalized approaches to the treatment of their patients. This scenario is technologically within our reach, and it is high time that we embraced it without losing sight of the privacy aspects.

Professor Magnus Ingelman-Sundberg performed research into the interaction between medicines and processes in the human body and the possibility of intervening accurately in that inter-action: the pharmacogenetics profession. He writes: "We face a future in which drugs are developed to a much greater extent for specific subpopulations, with genomic and other biomark-ers helping to provide more efficient personalized therapy. The principle of personalized therapy has been successful in increas-ing survival in, for example, patients with different forms of leu-kemia, where few new drugs have been developed but where treatment with existing drugs has been more personalized."[3] That individualization or personalization of medication is important progress. For example, a medicine for lung cancer will not work for most people but it works extremely well for 10 % of people. The difference is in the genes. People for whom the medicine works have a certain gene variant that the other 90 % do not. That knowledge of genetic biomarkers makes it possible to treat people accurately and increases the chances of success. Getting from 10 % efficacy on 100 % of a population to 100 % efficacy on 10 % of a population seems a small step at first glance. However, if you consider the consequences, you can see that it is actually a minor revolution.

These examples show that enormous gains are possible if we can develop more tailor-made solutions and that, in order to achieve these gains, it is often not even necessary to discover new active substances or to develop new health care methods. We 'only' need to have a clearer understanding of the data.

Unravelling and Interpreting DNA

Unravelling the human genome is essential to this process. It can give us insight into the diseases that can happen to a person and his or her sensitivity to medication. In essence, this unravelling is also 'just' an example of data analysis.

The developments in this area are fascinating. Craig Venter's Human Genome Project made it possible for the first time, at its conclusion in 2003, to map a complete human genome. At the time, the cost was $1 billion and it took 13 years. Since then, however, the race has had a central purpose: to offer an individual genome analysis—sequencing the three billion letters that determine our genomes—at a cost of under $1,000.

Such an acceptable cost level could bring it within reach of mass distribution within the medical world and would contribute to radical tailor-made medical treatments. That point seems to have been reached in 2013 by the American company Illumina that MIT ranked at number one in the list of smartest companies in 2014.[4] The price of such a genome sequencing system is still around $10 million, however, the cost price per analysis—meaning per person—is said to be below the magical threshold of $1,000.[5] The price decrease is happening much quicker than

expected, and is of course driven by the rapidly increasing calculative power of computers. In comparison, the price of a fully
unraveled DNA sequence was still $350,000 in 2008.

It is clear that giant leaps are being made in this field and the
result is that in the short term, we can get a much better insight
into someone's personal health and (if necessary) his or her
treatment. That improved insight is still the greatest challenge
however, says Illumina's CEO, Jay Flatley, in the article accompanying the MIT ranking. While unravelling DNA is currently possible, interpreting DNA must be the next step: "It's one thing to
say, 'Here's the genetic variation.' It's another to say, 'Here's what
the variation means." Our insight into the genome will improve
during the coming years. It will not cure all our ailments, but it is
definitely a breakthrough.

From Health care to Self-care

Another key development is that we increasingly practice Do It
Yourself (DIY) in health care, which may unburden professionals.
For example, the American company Theranos promises nothing less than a revolution in blood analysis. Its founder Elizabeth
Holmes wanted to eliminate the burdensome process of drawing
tubes of blood, transporting them to a phlebotomy laboratory
and waiting weeks for results. Therefore, she developed a technology that allows people to draw a drop of blood themselves in
a 'nanotainer'; they receive the results by return mail after analysis by Theranos. This methodology is a way to combine at least
a thousand different tests, completely painlessly and much more
efficiently than the current practice.[6]

Holmes' motivation (incidentally, she dropped out of Stanford
University) can be found in the possibilities of preventing health
problems instead of addressing them afterwards. What frustrates
her is that people still use weighing scales as an indicator of their
health, which is hopelessly outdated. "What's really exciting is

when you can begin to see changes in your lifestyle reflected in your blood data. With some diseases, like type 2 diabetes, if people are alerted early they can take steps to avert getting sick. By testing, you can start to understand your body, understand yourself, change your diet, change your lifestyle, and begin to change your life."[7]

Another example is a smartphone tool developed by a group of scientists at Columbia University in New York which can screen people for HIV and syphilis. Within 15 minutes, the tool provides a diagnosis based on a drop of blood and, in principle, replaces all the mechanical, optical and electronic functions of laboratory research. The cost of such a mobile lab: around €30. Its application is useful for example in African countries. In the words of the researchers: "By increasing detection of syphilis infections, we might be able to reduce deaths by 10-fold. And for large-scale screening where the dongle's high sensitivity with few false negatives is critical, we might be able to scale up HIV testing at the community level with immediate antiretroviral therapy that could nearly stop HIV transmissions and approach elimination of this devastating disease."[8]

Google is also keeping an eye on these developments, as demonstrated by a joint project between Novartis and Google's parent company Alphabet to develop a smart contact lens with a built-in chip that helps diabetes patients to continuously monitor their glucose levels based on the fluid in their eyes. An explanation by the researchers Brian Otis and Babak Parviz shows the origin of this very ambitious project—projects such as this are called 'moon shots' by Google—"Many people I've talked to say managing their diabetes is like having a part-time job. Glucose levels change frequently with normal activity like exercising or eating, or even sweating. Sudden spikes or precipitous drops are dangerous and not uncommon, requiring round-the-clock monitoring."[9] Google noted, however, that it will take years before such a product is ready for the market.

The Consumer Is in the Driver's Seat

The common thread between these concepts is that the consumer is in the driving seat. Before we enter the medical professional's waiting room, we have already informed ourselves extensively on the Internet. This is significantly altering the relationship between the doctor and the patient. It is therefore inevitable that the doctor will play a different role and evolve from an 'all-knowing god' into someone that guides patients in making the right decisions.[10]

This is a wider trend that is replicated across the entire care sector: consumers are increasingly fulfilling a role in their own health care process. A clear example of this is in elderly care where domotics, or the integration of technologies and services into our homes, allows us to automatically control lights, heating or devices and has seen a strong increase in recent years. With the help of technological tools 'care at a distance' is increasingly offered, allowing people to live independently for longer. This increases their sense of self-worth and it decreases care costs. This may entail a simple alarm button that a person carries on his or her body, but also more intelligent technology that monitors through motion and other sensors whether and when a person changes their regular routine, with or without the help of a smartphone. Sensors also make it possible to monitor for example whether a person is taking medication according to schedule, or whether a wound is healing correctly.[11] It is a great example of how the consumer, even if infirm, is increasingly in the driver's seat.

Quantified Self

These changes apply not only to people that require care, but also to the group of people that do not (yet) need care. In that respect, we cannot ignore the quantified self. This term, as mentioned in the introduction, refers to the fact that more and more people are so curious about the influence of something on their lives that they want to measure that influence. They want to get more insight into what they eat, how much they move, how deeply they sleep, how many hours they work, how their heartbeat changes with physical effort. They themselves determine how far they go—they can, for example, record how much chocolate or coffee they consume. Through various apps on a smartphone, they can simply record information on their own behavior, health, and lifestyle and choose whether they want to be motivated through games to reach the next fitness level, and therefore achieve improved health and wellbeing.

The human body has, therefore, become a quantifiable object. That 'measureable human' has a lot of potential for the future, especially because of the fun element that can be added to make it a powerful motivator for behavioral change. You can monitor how your results this week compare to last week, but also how they compare to other people with a comparable physical profile. If you want, you can even compete against them.

To many, this may sound like an unrealistic scenario. However, as *The Economist* said in the last words of the article *Counting Every Moment*: "Self-tracking may look geeky now, but the same was once true of e-mail. And what geeks do today, the rest of us end up doing tomorrow."[12] The rise of the quantified self seems unstoppable and seems to be becoming an essential part of our lives—for those who want it, at least. Many companies are catching on to the trend by developing wearable gadgets. It started with the smartphone itself, which makes it possible to count one's steps and to keep track of calories burned to promote a healthier lifestyle. More recently, various wearables have become available such as Fitbit, Jawbone and the Samsung Gear Fit.

Apple announced the Health Kit[13] and therefore clearly sees potential in the health monitoring market. The previously mentioned possibility of monitoring blood levels on a daily basis is one of the first highlights of this trend, because it opens up the possibility of matching lifestyle data to relevant blood levels.

You can probably guess the next step: the technology to gather information about our body will be placed on or in our body. Moreover, technology-wise, it is a piece of cake to upload the data on blood pressure, glucose levels, or iron levels automatically to the cloud every day. When we all start doing this, we not only become better-informed people; we will probably also demand that health professionals use the data collected to make better decisions. But more about this later in this chapter.

Insight into Insurance Products

Due to the increased measurability of our behavior and our health, a very fundamental discussion about changing (opinions on) social solidarity is starting within the health care industry. Since we know more and more about ourselves as individuals, we also know more about the risk of getting sick, being involved in an (industrial) accident or having other problems. We know exactly how our individual risk level deviates from the average. Up to now, a health insurer established a health insurance premium based on that average. That old concept of solidarity, which is the basis of all insurances, is however no longer self-evident.

Transparency Increases, Solidarity Decreases

The consequences for health insurers are profound. Every insurance company is built on the premise of imperfect information. One insured person runs a higher risk of getting into a car accident or getting cancer than another does. The 'lucky ones'—those with a low risk—are not aware that they have a lower than average risk and are in the same collective group of insured persons as those with a higher risk. In a certain sense, the lucky ones pay for the bad luck of others. One of the developments that the new information society is bringing however is that we are obtaining increasingly better insights into (individual) risks.

This also endangers the future business of insurers, in both the life and non-life sectors. The advances in sensors that we mentioned play a significant role in this respect because they are contributing to this growing insight into our individual health risks. As individuals, we are obtaining an increasingly clear picture of what exactly we're getting for the insurance premiums we pay. We can already see transparency in the industry increasing. That transparency is good, but it is also attacking the root of the current insurance business model. Because the more transparent the range of insurance products becomes, the more discussion we will have on the coverage and costs of insurance. Because we see exactly what we get for our money, what insurers make and the extent to which we pay for our neighbors' (higher) risks. In short, we see exactly how much that much acclaimed solidarity with others is costing us.

One of the results of these developments is a major sea change for insurers. In the past, people were unaware of this shared risk; we had to participate in a collective system, we had no other choice. Now we have more freedom and it will become clear how much intrinsic solidarity we have with others. In this new world, individuals will be able to define exactly what risks they are willing to take, with what other (groups of) people they feel solidarity and what form that solidarity takes.

One of the possible effects is that solidarity will happen more in defined groups, which are also called 'affinity groups' in the insurance industry. Smokers, for example, may decide to unite as a group and contribute to each other's health care if a growing cohort of non-smokers refuses to participate in collective insurance with smokers.

The new information society results in fundamentally different principles and difficult considerations. The position of a non-smoker who chooses not to pay for the health care costs of a chain smoker is ethically still easy to argue; the smoker can, of course, change his or her behavior. However, an increased premium for someone not as lucky as others with his or her DNA profile, and therefore with an above-average risk of getting cancer, does not fit so easily into our picture of a caring society—because no one gets to choose their DNA profile. Between these two extreme examples, there is a world of grey, in which judgments are harder to make.

In this respect, a comparison can be made with the example that we mentioned earlier in this book of Snapshot, the device that records the behavior—speed, braking behavior, distance, driving after dark—of drivers and makes it possible to calculate a tailored insurance premium based on that profile. This is in fact much fairer than using an average and, more to the point, it promotes safer road behavior, because each insured person pays for their own level of risk and not for that of a person known to be prone to accidents. In such a case, it is very clear how the behavior and the premium are connected. In the world of health insurers however it isn't always that easy. To what extent can you, as an individual, influence your own health?

There are some complicated dilemmas. However the development of new views about solidarity cannot be denied. It's a fact that a news reporter faces different risks than a construction worker, and an athletic person will require different health care than a confirmed couch potato. Since these differences are becoming increasingly easy to measure, it is quite conceivable that the number of new social networks and affinity groups that want to band together will increase. This may vary from people living in the same neighborhood, to patient associations, unions, friends, colleagues, or other groups. These people will be willing to share risks and help members who need it, with money or other support.

Insurers that facilitate this will, in a certain sense, be returning to the past where, as described previously, the concept of solidarity applied to a group of farmers who agreed to mutually support each other in case of a calamity. This very simple concept can be transplanted to modern times and improved with modern technology based on Big Data.

Do It Yourself (DIY) Society

Institutions—the insurer in this case—are being put to one side in the DIY society that is forming. This gives people a tremendous freedom to set up collectives themselves and to abolish the old institutions. Take a group of neighborhood residents that buy a windmill together and no longer need the energy company. Or look at how groups of entrepreneurs fund each other's ideas through crowdfunding and therefore no longer need banks. In short, we are doing more for ourselves and breakthroughs brought about by Big Data could be the ultimate push in that direction.

Will Health Insurers Get a New Role?

Everything we have written in this chapter up to this point can be summarized in one sentence. By using data intelligently, we can enjoy longer, healthier lives, especially if we play an active role ourselves.

That is a nice outlook. Is it not highly logical, therefore, that all parties in the health care sector should help us to achieve this? The answer is yes, with one essential precondition. The system needs to include the appropriate financial incentives to invest in prevention.

Health care statistics leave no room for misunderstanding. The majority of health care costs are incurred in the final years of our life. When we become capable of living, on average, five years longer in good health, these costs remain the same or even increase. The curve goes up at the end of life and only moves somewhat in time. When viewed from this angle, there is no financial incentive for a health insurer to keep us healthy for longer. There is, however, a social gain in living longer in good health. The challenge is to set up the health care system such that there are incentives to invest in prevention.

Can we change the focus in the health care chain completely by no longer insuring our health care, but our health? This is a radical change in how we think about health care. At the moment, everything is mainly focused on controlling—or rather reducing—health care costs, including because of political pressure. A popular creed among purchasers, however, is that managing based on costs results in lower quality, while managing based on quality results in lower costs. That wisdom should be central to the transformation of health care. We are facing the challenge of investing in the health of people such that every euro invested—whether in prevention or in actual medical care—results in healthier people and in more vibrant organizations and lower social costs. Health care costs are only part of that jigsaw.

As we said, the current system does not encourage (investments in) prevention. As is often the case, history provides valuable insights, in this case from Chinese medicine. That has not only been based for thousands of years on the prevention of health problems, but for centuries it has also had an extensive system that rewards doctors for the health of their patient populations. Even today, Chinese general practitioners receive a basic salary that they can increase if they are capable of improving the health of their patients. That simple—and fair—motivation could very well be applied to our health care system, especially since we have increasingly more good data. The possible result is that we will be capable of taking prevention seriously; we are stimulated to map the individual impact of treatments on health and are therefore capable of better controlling the overall health-related costs, not just health care costs. And the most important advantage is that we can live longer in good health, on average.

Privacy: A Great Challenge

The challenge to the entire health care chain is not a small one. As individuals, we expect radical tailor-made medical treatments instead of a scattergun approach; as the individualization continues, we will probably only pay health care premiums in the future for risks that apply to ourselves; and we want access to the most advanced—and expensive—treatments should we need them.

However, at the same time, we do not want to hand over our personal data, especially not where our health is concerned. That is partly because we do not trust health insurers to use that data to look after our interests. One of the key themes when we want to initiate change is: under what conditions are we willing to share personal information about our health—clinical data, preferably combined with data on our lifestyle and diet—for the purpose of medical science? In Europe, in particular, there have been concerns for many years with regard to privacy and possible abuse. The concerns were confirmed by the discovery by the British newspaper The Telegraph that information on 47 million patients was sold by the National Health Service (NHS) to insurers, who were then able to determine the patients' premiums based on that information. It is hard to see how that transaction could contribute much to the health of patients[14]—even though that is exactly what health insurers should be aiming for.

The question is whether these privacy concerns will continue to be as important when the medical sector is capable of clearly showing the benefits to patients that are willing to share data thereby making radical tailor-made treatments possible. Society's attitude may change drastically once people realize what the advantages are, both for their own treatments and for the development of science and with that the contribution to the health of others. We may even completely change attitudes towards the privacy aspects surrounding electronic patient files. If you are not prepared to share data, you are being selfish, because you are frustrating the development of better medical treatments for your fellow human beings. The challenge? Sharing data without endangering privacy.

Technologically, this is possible. Various parties are developing advanced solutions that respond to privacy requirements resulting from legislation. This technology makes it possible for the party analyzing the data—whether it is a health care provider or health care insurer—to use the data on an individual level, but without being able to trace it back to a person. When required, decoupling—anonymization—can be done by a third party. In this case, an insurer, through that third party, can provide the insured person with personal advice based on Big Data, without knowing that person's identity.

A group of scientists at the Massachusetts Institute of Technology led by Alex 'Sandy' Pentland is developing a 'New Deal on Data', with open source tools for controlling, storing and auditing flows of personal data, collectively known as openPDS. This technology aims to mitigate privacy risks. The basic premise is based on three principles[15]: "you have the right to possess your data, to control how it is used, and to destroy or distribute it as you see fit". Pentland's team is developing 'trust networks' in which people can take back the control of their personal data. The success of these projects is very important. The use of data is extremely promising, especially in health care, but the progress may be significantly hindered if we do not provide guarantees around individual privacy rights and control.

Transforming Doctors into Data Scientists

When the Big Data revolution in health care described here actually happens, it will result in a (drastic) change in how care providers operate. The doctor of the future can no longer arrive at a proper diagnosis based only on his or her own knowledge and experience, but will have to be supported by data analysis with the proper technological tools. The generalists will have to develop into medical data scientists that also act as health care agents. For this purpose, they require mathematical (for example statistical) knowledge in order to better predict health problems using patterns in data.

Doctors are increasingly more supported by 'Watson-like' systems that allow for holistic diagnoses. Watson is IBM's supercomputer that became famous in 2013 when the system was able to beat two human candidates in the TV show *Jeopardy*—the best two candidates from the show's history. The computer can interpret a question asked in a spoken language and answer after a lightning-fast search through numerous information sources. The software developed goes beyond the usual artificial intelligence: Watson is also for example capable of recognizing irony and riddles.[16] Watson's potential became clearer in recent years. It is now capable of competing with human head chefs by creating tasty recipes, a process that requires much creativity. This cognitive cooking is a taste of what IBM is really up to; the company sees Watson as the smart assistant for many sectors, from retail and the financial industry to health care.

It is clear that IBM sees (commercial) potential for Watson in the health care market and has already demonstrated how Watson together with a health care professional can make diagnoses. Based on large amounts of data, Watson generates possible diagnosis paths which the health care professional can research in order to make a well-founded opinion on the best possible treatment.[17] Integrating such systems into the health care chain in order to identify connections much more accurately than before is one of the great challenges facing the medical sector today.

Of course, IBM is not the only organization that sees a future in health care that is more data-based. That is why the XPrize Foundation, a non-profit organization, and Qualcomm, an American telecommunications giant, organized an international design competition with a simple idea: design a mobile solution that can make the same or even better diagnoses than a panel of doctors. The Qualcomm Tricorder XPrize—named after a gadget from Star Trek—offered millions for the best portable, wireless device that fits the palm of your hand and simply scans your body and tells you how healthy you are. No doctor, no medical manual, no MRI scan, or X-ray required.

In order to meet the promises in health care made by Big Data, it is necessary to combine various information sources and therefore to achieve a collaboration between various parties. Work is done on many fronts, including in the Pittsburg Health Data Alliance—which aims to take data from various sources (such as medical and insurance records, wearable sensors, genetic data and even social media use) to draw a comprehensive picture of the patient as an individual, in order to offer a tailored health care package. The way they put it is that the health care field generates an enormous amount of data every day. There is a need, and an opportunity, to mine this data and provide it to the medical researchers and practitioners who can put it to work in real life to benefit real people. Many organizations can do parts of this process, but none of them is equipped to begin with raw data, develop an idea, and move that idea directly into a practice setting. This alliance brings relevant parties together to do this for the first time.

Above all, one thing is becoming clear: it is almost inevitable that role of the general care provider—the general practitioner or the internist—is changing. It will increasingly be about interpreting and combining data instead of drawing conclusions based on a limited data set.

There is a certain renaissance of past values here, but in a modernized format. In the past, every doctor made his own choices based on his medical knowledge, experience and knowledge of the patient. Your general practitioner knew (the data on) your lifestyle without having to glance at your file.

That 'artisan' approach made way for far-reaching standardization via protocols and guidelines to which all doctors have to adhere. There are pros and cons to this development. The intention was good: this method makes it possible for knowledge of successful and less successful treatments to be shared easily, allowing doctors to learn from each other and to avoid having to reinvent the wheel. The big disadvantage, however, is that there are hardly any possibilities left to personalize treatments.

We argued previously that such a tailor-made diagnosis could drive a big leap in the quality of health care. The good news is that the possibilities in this respect are near at hand because of Big Data. We are therefore facing the challenge of making the health care sector deal differently with our data so that standardization and personalization will be seamlessly joined together.

Data analysis potentially offers an enormous wealth of possibilities to increase our medical knowledge. We know, for example, that people who eat too much and too fat, have a bigger chance of developing cardiovascular diseases. And that smokers die more often of lung cancer than non-smokers. And that drinking alcohol during pregnancy increases the chance of birth defects. However, discovering these connections was more of a coincidence than the result of focused research. We can now focus in on these types of connections since we have much more data at our fingertips.

A doctor cannot do that for various reasons. First, he is not a statistician; secondly, he only sees a limited number of patients; and third, the human brain is not capable of making complex connections between various variables on the scale required by medical research. However, this can be done through data analysis under certain conditions. An American company active in this field is Enlitic. Enlitic applies deep learning techniques to analyze data and to formulate medical diagnoses. The objective is not to make doctors obsolete, but to give them a tool to do their work better. According to its founder, the promise it makes goes beyond Watson: "Watson mimics medical science in the pursuit of creating an artificial super doctor that knows more than any single doctor could ever learn. But Enlitic could potentially make new discoveries by uncovering previously unnoticed patterns in the data."[18]

Pharmaceutical Industry

The extent of the challenge we face to change health care is clearly visible in the pharmaceutical industry. We do not want pills that work for the average population; we want pills that are exactly right for us as individuals. Based on large quantities of data—clinical data, genetic data, and data on our personal lifestyles—we expect targeted medication. It is a fact that individual characteristics alter the effectiveness of a medicine and it is time to recognize this.

The development of such personalized medicines requires not only another way of dealing with data, but also a completely different business model. In the current model, it can often take 10 years—and sometimes up to 17—of scientific research to arrive at applied use of medicines. First, significant investment is required in the development of a generally applicable medicine and these costs have to be recouped through large volumes of sales. This model does not fit personalized medicines: both the development time and the idea of general applicability are unacceptable with the current level of technology. The question is how we can transform this sector into a responsive, flexible, and customer-focused industry, while maintaining the diligence that is so very important.

People Want This

Can people themselves take the lead in this transition and even get the transition going? It is possible when we think about concepts such as the quantified self. If the health care chain, through lack of flexibility in its processes, finds it difficult to react to our changing demands, we can shape that change through crowdsourcing.

To underline this, we can look at the model of the American startup Exogen Bio. This company is building a business model based on mapping the daily damage done to our DNA.[19] The damage to our DNA is a continuous process, caused by factors including what we eat and breathe, and can result in health problems. In terms of science, we have very little insight into what exactly that damage means in individual cases. Exogen Bio wants to change this by scanning enormous amounts of information for patterns and thereby preventing health problems. The idea is for thousands of participants in the program to submit drops of blood using a special kit. Through this crowdsourcing, the company will—hopefully—obtain valuable insights. It also hopes to be able to warn its users when their DNA damage shows patterns that have been found to be dangerous after mathematical analysis and in comparison with the other users' data.

Some nuances are required where this type of initiative is concerned. Crowdsourcing is a very powerful tool to collect the insights and information of a large group of people, but it is at least some sort of concern when the analysis of the information about a person's health is done (in part) by an (untrained) person himself. The input of specialists, who draw conclusions based on good qualitative analysis of possible treatments or medication, should be ensured. Americans have a great expression for it: "A fool with a tool is still a fool." We have little insight into the medical expertise and competencies in the area of data analyses where initiatives such as that of Exogen Bio are concerned. This can result in troubling situations. Where this type of initiative is concerned, it is important that the appropriate level of expertise is present, especially because not all of the organizations concerned are affiliated with a health care institution.

In this specific situation, it may be wiser to limit the input of crowdsourcing to collecting information and to leave the analysis to the specialists. This is the case not only for health care, but also for a wider area. We will discuss this in Chap. 7.

A great example of having proper assurances can be found in the British charity Cancer Research. This organization uses crowdsourcing to improve medical treatments by using a game called

Play to Cure: Genes in Space. This game is played on a smart-phone and is based on large databases with information on the activity of genes. The institute outsources the manual research work to thousands of players in the gaming community. The assignment is simple: fly a spaceship through a field of asteroids. At the same time, you are mapping a pattern that is represent-ative of the variations in gene activity and that in turn provides information on the development of cancer.[20] The example shows how people can play a part in processing data and how that can contribute to improved health care while the analysis of content is done by specialists.

Such examples, however, are scarce in the medical world. The difference with the commercial world is striking, where customer profiles have become generic, in order for companies to provide us with targeted information, services, and advertisements. In the medical world, however, such methods based on personal pro-files are still a long way off.

Finally

In this chapter, we have been preaching about nothing less than a health care revolution. A revolution in which we will give the quality of diagnoses and treatments a significant push by being willing to share our data. A revolution in which we will invest in prevention instead of discussing the costs of health care treat-ments. A revolution in which data analysis becomes an important task for all medical professionals.

But Is that Realistic?

First, such a drastic change is not easy and will take time. One significant factor is that no single party within the sphere of health care is dictating the rules. If we want to alter the course of the entire system, the vision, willingness and perseverance of all the actors within that system is required.

The answer given by Sergey Brin, co-founder of Google, to the question whether he would do business in the health care sector is typical of attitudes: "It's just a painful business to be in. [...] I think the regulatory burden in the US is so high that I think it would dissuade a lot of entrepreneurs."[21]

Second, we have to ask whether such a transformation is, in fact, unattainable in light of the discussions about continuously increasing health care costs. Is personalized medication completely unaffordable and therefore unattainable? Due to the increased demands and expectations of society, there will undoubtedly be more and more initiatives to provide affordable tailor-made solutions. However, it will take some time before we all can profit from this, even though part of the investments will be offset by cost savings.

The previously mentioned professor Ingelman-Sundberg is very clear about that: according to him, at least 7 % of hospital admissions are the result of the side effects of pharmaceuticals. For people of over 70, that figures rises to 32 %. His conclusion: "The costs of treating side effects are almost as high as the total treatment costs." In his report, he estimates that in the Netherlands—a small country with a population of 17 million people—there are between 10,500 and 16,000 possibly avoidable admissions annually. Therefore, huge savings—and healthier people—are possible in the long term.

It is about time that we starting aiming straight for the bullseye. When we collect data on our bodies and our behavior—and also share the resulting insights—we can achieve tailor-made solutions in health care. With the help of new technological developments, combined with Big Data concepts, a few of which we discussed in this chapter, this will hopefully lead to success in an affordable and generally accessible manner.

Chapter 5
Data Stimuli for a Better World

Now that the existing world is reaching its limits, new equilibria are being created

Against All Expectations

In 1990, the New York City authorities closed 42nd Street to traf-fic. By doing so, they took an important route out of the city's road network. Many thought this would result in an enormous mess with many traffic jams because it was such an important thoroughfare. The opposite was true however. The traffic flow in the entire area improved.[1]

As shown in previous chapters, Big Data is an important basis on which to build a new information society. Bringing this one step further brings us to the question whether we can create a more sustainable world with the help of Big Data. We will address this conundrum by laying a theoretical foundation to weave social developments—as described in the previous chapters—with the possibilities offered by Big Data. Then, in the subsequent chapters, we will address the assurances required to make that link effective and efficient.

It Is Time for Sustainability

A sustainable world is a world that meets the current needs of humanity without endangering the needs of future generations.[2] It is clear that sustainability is under great pressure. The global

© Atlantis Press and the author(s) 2016
S. Klous and N. Wielaard, *We are Big Data*,
DOI 10.2991/978-94-6239-183-3_5

population is growing rapidly and changes in consumption and lifestyle patterns result in an increased demand for raw materials instead of the required decrease. According to estimates by the World Wildlife Fund, we currently consume 50 % more natural resources than the Earth's ecosystem can replenish.[3]

These needs include, for example, our basic need for food, but also for energy. Fossil fuel supplies are slowly being depleted and the same applies to various essential raw materials. Climate change may already lead to problems in the coming decades. Finally, biodiversity is also seriously under pressure, with potentially disastrous consequences for our food supply.

The sustainability issue is a many-headed monster, but in essence it is also very simple: it is a scarcity problem. We use too many raw materials and consumables, we pump too many hazardous emissions into the atmosphere, we cannot produce sufficient food to feed the entire global population, and we are facing a growing divide between the haves and the have-nots. We could try to fix the problem or at least reduce it to manageable proportions if global leaders were to make binding agreements. However, that does not seem certain in the short term. The United Nations Climate Change Conference saw more than 190 nations gather in Paris in December 2015 and reach agreement for the first time to keep global temperature rises to 2 °C. However, this agreement is as yet tentative and only starts to address the challenges we are facing.

In recent decades, the solution was sought in behavioral changes. The reasoning behind this was that if we were aware of the consequences of our behavior, we would change that behavior. This is the reason that people organize events like Earth Hour, to make us aware of the excessive consumption of energy by the lights we use every day. Advertising is motivating us to purchase organic produce, to opt for sustainable energy sources, to work out more often to prevent obesity and to only drive our cars outside of rush hour. The message that we should consume less, set the thermostat a degree lower, or get used to less luxury is not at all popular and therefore also not very effective. This does not mean that we should stop creating awareness of the problems and try to contribute to behavioral change, but the impact of this message is not enough to adequately battle the many-headed monster. We should, therefore, also look for other solutions. We are convinced that Big Data, linked to new mechanisms, can help us in this respect.

It Is Time to Collaborate

An example of such a mechanism: visualize a living room, with a big pot of delicious-smelling soup on a table. Sitting around the table are skinny, sickly looking people. Spoons that are longer than their arms are attached to their arms. Therefore, they cannot bring the spoons to their mouths and they are slowly starving. Now visualize another room, with exactly the same setting, but with a big difference. The people are all well-fed and healthy. They are laughing and talking to each other. And they can do that because they have learned to feed each other with the long spoons.

This metaphor, from an interview with the former UN climate chief Yvo de Boer, is a vivid example of how we are capable of solving problems when we collaborate seamlessly instead of waiting for the other to act first.

When we use the right mechanisms—in this case the long spoons to feed each other—anything is possible. What does this have to do with Big Data? Everything. As we show hereafter, Big Data provides possibilities to design mechanisms that promote sustainability. Economic science can help us achieve that.

We do have to make sure however, that our understanding of the term 'the economy' is not limited—as is often the case in daily discourse—to everything related to money. In essence, economic science is about the needs of human beings, wherever they may be in the world, both the current and future generations, in terms of both time and space. One of the basic principles in economic thinking is that individuals and companies maximize their own value and through that also create a wealthy society.[4] Unfortunately, that principle is not always applicable to just any real-world system without considering the boundary conditions, which in essence has led to the sustainability crisis.

For example, take the commuter that gets into his car every day and joins the traffic jam on the way to work. The commuter's reasoning is mainly based on self-interest and he does not think about the delay that he is causing for the people that will join the queue of traffic behind him. The choices that we make are largely disconnected from the consequences of those choices for the people that come after us. This phenomenon is known as the tragedy of the commons[5] and plays a key role in the sustainability problem: maximizing the value for ourselves is taken to such an extreme that the societal value is ultimately diminished. Back to the example of the commuter: because of self-interest, the commuter chooses to hit the road at the time that best suits him, which also contributes to the collective traffic jam.

The Equilibrium of John Nash

A scene from the movie *A Beautiful Mind* (2001) provides a great illustration of the problem that keeps us imprisoned. Student John Nash—played by Russell Crowe—is in a bar with some friends. A group of beautiful young women comes in and is immediately the focus of the whole bar's attention. Five brunettes and a blonde woman, clearly attracting all the male attention, even that of the otherworldly Nash, who is really preoccupied with his thoughts of mathematical theories. The blonde woman is obviously charmed by Nash, and his friends cheer him on to take action.

Suddenly Nash has a brilliant thought and exclaims that Adam Smith was completely wrong. His friends don't understand what he's talking about. He explains that the famous economist in his theory on what is known as the 'invisible hand' of the market states that we maximize wealth by channeling self-interest. However, according to Nash, that theory can be thrown out. Because if everyone chases the blonde, they will be in each other's way, and no one will get her. And then? They will try the brunettes, who will also reject them, because no one likes to be second choice. Therefore, Nash says, none of them should go for the blonde—they will not be in each other's way and the girls will not feel insulted. That is the only way they will all get to take a girl home tonight.

Nash takes his papers and walks over to the blonde woman who clearly expects him to hit on her. However, he only whispers "thank you" in her ear, leaving her confused as he exits the bar. On reaching home, Nash starts writing his paper, which will win him a Nobel Prize many years later. The Nash equilibrium has since become an established concept among mathematicians.

A proper understanding of that state of equilibrium is essential when creating new equilibria in society, which has become necessary as we are now reaching the limits of our system, meaning that sustainability of that system is threatened. Until recently, (large) companies could achieve commercial success through increasingly aggressive expansion strategies. That time is past. The world is becoming more and more a corralled ecosystem, with jamming throughput and bottlenecks. This applies to both raw materials and consumables as well as economic growth and data, communications and information. The development of wealth and well-being is especially dependent on the question of whether we can limit the negative effects of the Nash equilibrium states in the system, in other words, whether we can think of something powerful enough to motivate each other to do things for the collective good.

The Nash equilibrium state is based on the assumption that each participant in a system (player) wants the optimum result for himself and does not collaborate with others. In other words, it is a non-cooperative equilibrium. Game theory teaches us that such a 'selfish' approach often does not result in the most efficient or socially beneficial state of the system—called the Pareto optimum.

What is more, if players all act purely from self-interest and do not collaborate, expanding the available resources may even lead to an unfavorable equilibrium. This is known as the Braess paradox. This argues that adding capacity to a network in which moving entities determine their paths individually, may result in a decrease of the general performance of the network. We mentioned at the beginning of this chapter a famous example that occurred in 1990 in New York City. Authorities closed 42nd Street and reduced the size of the traffic network for cars. Despite this restriction traffic flow improved. The question now is whether we, with the right insights into the effects, could introduce mechanisms that could change our behavior, like what happened in New York, and in doing so build a sustainable world.

It Is Time for New Habits

Are we capable of changing our behavior in the information society? And if so, where will we find the willpower to not touch the proverbial cupcakes because we know that we are too fat already?

It is not about willpower, but about learning new habits. We don't necessarily have to unlearn bad habits, but we do need to learn good ones, and the environment can play an important role in this respect. There was a reason that Michael Bloomberg, in his time as mayor of New York, developed a different vision on construction in the city: buildings had to be designed to encourage the use of stairs, allowing people to form a new habit. The question is, how we can use the Internet and connected computer systems to learn good habits?

Can using Big Data motivate us to learn new habits, or even force us to do so? Would it then be possible to introduce mechanisms that would motivate us, as individuals or businesses, to do the right thing, both from self-interest and collective interest points of view? Doesn't Big Data present us with possibilities we could only dream of before now?

The key to using those possibilities to reach a new, hoped-for equilibrium can be found in the 'mechanism design' theory. In 2007, the American economists Hurwicz, Maskin and Myerson received the Nobel Prize[7] for their ideas in this area, which is also known as the 'reverse game theory'. Mechanism design is about designing social constructs, procedures, incentives, and 'game situations' that bring us closer to established social objectives. The theory supposes that we as players will act in a manner that is for the good of society as a whole. Myerson describes[8] mechanism design as a tool to achieve economic or social objectives by addressing individual emotions, incentives and motives.

It requires a lot of imagination and fantasy to imagine where mechanism design may lead us, because for now it is mainly theory. However, it is possible to explain to a non-scientist audience what the results may be. In the Netherlands, Arnold Heertje, emeritus professor of economics at the University of Amsterdam, discussed it a couple of times; he also used some simple examples to explain the concept.[9]

One of those examples is of a mother that has a piece of cake and wants to divide it equally for her two children. Many parents can predict what will happen: both children will claim that they are entitled to the largest piece. Because they are the oldest and therefore have the largest stomach, or because they are the smallest and still need to grow. This conflict can be very easily solved by mechanism design: child A gets to cut, child B gets to pick the first piece. This leaves child A with only one strategy: to divide the cake equally. This is in his or her interest, but also in the interest of the other child.

To Force or Motivate Change

In the new information society, we can use intricate information from and for participants, which we did not have before, to solve such conflicts of interest. This provides us with a new arsenal of possibilities to align interests using well-chosen stimuli, thereby contributing to a better world. This may sound a bit too conceptual, but it is already occurring in places. A great example is a pilot project that is currently being run in Delft, the Netherlands, to improve traffic flow. Researchers at the Technical University there have developed a model that not only follows traffic flows at the level of individual cars or roads, but covers a large area, including outside the city proper.[10] By combining information on certain areas and aligning traffic lights to that information, traffic flow can be improved. The nice thing about the project is that an individual may possibly be forced to wait five minutes longer at access roads but still arrive home 10 min earlier than usual. The additional waiting time is perhaps counterintuitive, but there is no doubt that it contributes to both the individual's interest and the collective interest by improving the traffic flow. In this example, there is not even a choice: the system forces us through the use of traffic lights to change our behavior. We will address this further in Chap. 10.

Other options are possible—and are being developed—to force or motivate change and thereby promote sustainability. One of those is the rise of 'smart money', a subject in which Big Data also plays a central role.

Currently, our money is dumb. It has no characteristics, no preferences, and no preconditions. Back in the day, there was no other option. A physical coin or banknote only has one characteristic: the value written on it. Apart from that, we can do what we like with it. Money such as the Bitcoin, however, is made from data, and that makes it possible to add conditions to how it is used. This is one of the differentiating features of Bitcoin; it is possible to set a condition that, for example, payment for a car may only occur when legal ownership of the property has actually been transferred. As described earlier, this is a great way to make transactions without an intermediary reliable, but it has also triggered a movement towards smarter, programmable money, with payment conditions attached.

A prime example of this is a patent application by the auction site eBay for gift tokens that look a lot like Bitcoins in more ways than one. eBay—the owner of the payment platform PayPal—is capable of assigning various conditions to the digital credit.[11] For example, the credit has to be spent within a certain period; it can only be spent within certain countries; its value is €25 if spent on games and double that if spent on educational books. You don't need much imagination to go from here to dreaming up other criteria that programmable money could offer. An employer could attach a condition to 10 % of your salary that you spend it in a sustainable manner. Or block another 10 % in a savings account for five years to motivate you to spend your money wisely. Of course, this brings with it some major ethical issues, because such actions are too close to control for comfort. However, this does not diminish the fact that programmable money puts the potential to change the behavior of people and organizations within our grasp. It is a powerful tool to bring the ideas surrounding mechanism design from theory into practice.

Communications and Feedback

An essential part of these mechanism design examples is the effect of communications between the participants, who often all have different information. This is exactly where the link to the potential of the new information society can be found. First, because it makes it easier to map the consequences, positive or negative, of actions, which will enable more well-founded choices. Second, because Big Data makes feedback mechanisms possible that are much more intricate than previously feasible. We are capable of creating insights that are important at an individual level (not "what should society do?", but "what should I do?") and we can also provide feedback at an individual level (not traffic information on the radio, but tailor-made travel advice on your smartphone).

Part of the challenge lies in creating an accurate picture of the effects our actions have on ourselves, on the interests of others and on society as a whole. By mapping these negative or positive consequences, we may be able to create the proper stimuli for a more sustainable equilibrium. This is already happening on a small scale. In 2009, the car manufacturer Volkswagen launched a promotion to change people's behavior by giving them fun stimuli, including turning the stairs in a subway station into a piano, to motivate people to exercise. Every time someone stepped on a stair, it made the sound of a piano key. The number of people taking the stairs increased by 66 %. This experiment shows that you can also influence behavior through fun activities and of how we can achieve social objectives despite the personal (short-term) interests of people involved.

Jeremy Rifkin, an authority in the field of sustainable ecosystems and writer of the bestseller *The Third Industrial Revolution*, says that this type of mechanism will result in improvement of energy efficiency, a dramatic increase of productivity and finally to a sustainable circular economy. He mentions the 'collaborative age' and sketches how the Internet of Things is already developing fast, including because of the impetus given by big companies such as GE, Cisco, IBM, Philips and Siemens. This creates an intelligent infrastructure that enormously increases the potential for the previously mentioned feedback to participants and therefore offers room to bring the mechanism design theory of Myerson, Hurwicz and Maskin into play. This is made possible because the Internet creates a global network in which everyone can design apps to make their own lives more comfortable and/or to combat social problems.

The experiment in Delft, the Netherlands, to control traffic flows to serve both our self-interest and the collective interest, is inspiring. However, it is only a start. In the new information society, we will have intricate data that will create solutions that were previously not within reach. We are not sure where this will lead. However, it also holds out the promise of being able to 'close the loops' macroscopically, creating a circular society in which sustainability problems are solved.

Is Realizing a New Equilibrium Obtainable and Stable?

The fundamental requirements and scientific insights to achieve a new equilibrium are present. Nevertheless, many challenges will have to be overcome before we are able to bring such daydream scenarios into practice. Because what are the right building stones to achieve an acceptable new information society? In other words, how can we achieve the desired optimum in a controlled manner and how will we subsequently maintain that equilibrium?

The 2010 Wall Street flash crash is a clear example of an information ecosystem that becomes unstable if only a (colored) part of the information is available. At that time, the Dow Jones Index

fell by 10 % in the space of a few minutes before returning immediately to its previous level. The high frequency traders' automated systems took buy and sell decisions based on very limited
information, creating an unstable market.

Will the equilibrium that will be created in the new information
society under the influence of mechanism design be more or less
sensitive to these so-called tipping points than our current economic system? A tipping point is the critical point in an evolving
situation that leads to a new equilibrium. The term has become
known to a wider audience because of the book *The Tipping
Point: How Little Things Can Make a Big Difference* by Malcolm
Gladwell. In this book, he examines the tipping point effect in
social systems. He shows how small differences can have big consequences in society and talks about 'social epidemics'. Just as
one sick person can start a flu epidemic, one right move can also
make all the difference.

The idea of the tipping point has existed much longer than
that and was also used by the scientist Thomas C. Schelling. As
far back as 1978, he described in his book *Micromotives and
Macrobehavior* how a small difference could have an enormous
impact on race segregation in a neighborhood. Tipping points
also occur in other areas. In the field of climate change, a well-
known workshop was held in the British Embassy in Berlin,
October 2005, titled *Tipping Points in the Earth's Climate System*.
At that conference, an international team of experts tried to identify relevant tipping points in the Earth's climate system.[12]

In order to study the influence of information on the stability
of a data-driven system, we need to look again at game theory.
In 1994, John Harsanyi together with John Nash and Reinhard
Selten received the Nobel Prize. One of his key contributions was
the analysis of incomplete information in game theory. A resulting insight is that the stability of a system depends on the com-
pleteness of information. When players are fully informed of the
consequences of their choices on their own situation and that of

others, the stability of the system increases. This calls for radical transparency by companies and governments regarding their activities. In daily practice, we have seen the impact of this mechanism on the complex financial derivatives that were the basis of the financial crisis in 2007. Would consumers have followed the same strategy of taking on (sub-prime) mortgages if that information had been brought to their attention? Probably not. As a result, the equilibrium was not stable. Furthermore, during the crisis it became clear that the products had become so complicated that even financial institutions did not understand the damage to their portfolios. Transparency has become one of the central principles of the restructuring of the financial sector in order to avoid a repeat of such disastrous scenarios.

Transparency, therefore, plays an important role in achieving stability in ecosystems. Game theory uses the term 'ex-post stability analysis'. This analysis reviews how satisfied players would be with their results if a full set of information were to be made available to them. When there are many players that could have performed much better by opting a different strategy, the equilibrium is less stable.

Does the rise of the new information society, as described in previous chapters, lead to sufficient transparency to maintain a stable equilibrium? The image is diffuse. On the one hand, there is increasingly more symmetry of information between participants since all the information becomes available and becomes increasingly easier to access. On the other hand, that omnipresence of reliable information is perhaps an illusion. Although we live in a world in which all information seems to be available, in reality we only have a small window on that world, because our information is increasingly tailor-made for us. This phenomenon, known as the filter bubble (which we will discuss in Chap. 8) means that we live more and more in our own information ghettos. The 'colored' information that we have to deal with may even result in additional tipping points being reached.

Make All Costs Visible

Nevertheless, we are convinced that mechanism design can at least contribute to solving the sustainability problem. One of the conditions in this respect is that we need to have a clearer insight into the undesired (side) effects of our actions. We don't have to respond only to idealistic motivations, because the sustainability crisis will (at least in time) also become an economic crisis. Attempts are being made to introduce a new standard for economic growth. Actually, we need to embrace an entirely different language when we talk about the economy. Terms such as productivity and growth are too limited. The OECD talks instead about 'green growth'. Green growth means fostering economic growth and development, while ensuring that natural assets continue to provide the resources and environmental services on which our well-being relies.[13] However, in most political discussions, the economy is still reduced to discussions about money.

It is because of this attitude that the area of sustainability is in such a state: it is hard to put a price tag on a fish killed by pollution or acres of forest lost through illegal logging. The damage done by companies and people to the environment by overusing scarce resources is not translated into costs for themselves, but is passed on to the next generation instead. These hidden costs—in economic jargon 'externalities'—rose by 50 % from $566 billion per year in 2002 to $854 billion in 2010 according to a global study in 2010.[14] In other words, every year we are 'stealing' $854 billion from future generations. Those costs are not visible in the annual accounts of companies, although there are some early adopters who are voluntarily experimenting with making these externalities visible under the title 'true value' or 'true cost'.

Peter Bakker, president of the World Business Council for Sustainable Development, is of the opinion that "accountants will save the world" if reporting is significantly changed.

His reasoning is that when a company's reporting shows its impact on society and also establishes a link with value creation, a lot of change will start to occur. Companies will be motivated by their numbers to do the right thing. A small vanguard is very active in this respect. The sports brand Puma, for example, is showing that it is possible to calculate the true cost of a company's impact on society and make changes to its operations and supply chain based on that.

To give an insight into the results of this exercise, the environmental impact of the conventional Puma Suede sneaker amounted to €4.29 per pair, while that of the new biodegradable InCycle Basket sneaker was only €2.95—which equates to around a third less environmental damage across the product lifecycle.[15] The Danish pharmaceutical company Novo Nordisk does something similar. The national railroad company in the Netherlands mentioned an 'alternative profit calculation' for the first time in its 2013 annual report. The company wanted to show its 'true value'. What was the result? All train trips combined 'cost' the environment €59 million in 2013. On the other hand, there was also an environmental profit because at least some of the train passengers would have taken a (more polluting) car if there had been no train. The balance? €1 million "positive".

The Rise of Integrated Reporting

Of course, the underlying calculations are complicated, and it is not a perfect system. However, it is a key development that is increasingly being adopted and that, with the rise of Big Data, may result in increasingly sharper insights. More and more companies are embracing a type of integrated reporting. The essence is that a connection is made in the annual report between how an organization reacts strategically to broad social themes and the associated (financial) value creation. This development is getting up to speed; new tools and data analysis concepts can contribute to that. The objective is a better and more refined insight into the consequences of our actions in a system that is reaching its limits.

Accountants could be at the bottom of such a system and thereby initiate a fundamental social change—although not many people will associate them directly with such grandiose deeds. The Italian invention in around 1300 of a system of double-entry bookkeeping with debit and credit sides, which is still in use nowadays, inspired enthusiasm and discipline and was the basis for modern day capitalism. Jacob Soll, author of *The Reckoning: Financial Accountability and the Rise and Fall of Nations* says "Good accounting practices have produced the levels of trust necessary to fund stable governments and vital capitalist societies, and poor [...] have led to financial chaos, economic crimes, civil unrest, and worse."[16] Since we also want (or have) to account for the costs of sustainability—and in doing so introduce what is, in reality, a new form of capitalism—accountants could achieve immortality through the rigorous re-invention of the system of double-entry bookkeeping and thereby contribute to a better equilibrium, fed by the possibilities Big Data offers.

We need to realize, however, that there is more to this new equilibrium than just sketching its outline, but also the manner in which we will grow towards it. A purely theoretical discussion helps no one. Suppose that the British government decides tomorrow that it would be better to be in line with the European mainland and, from next week, traffic in the UK will drive on the right-hand side of the road. On paper, that may appear to be a good idea. However, the transition to the new situation would require significant adjustments that may not be realistic in practice.

This applies to every change, including the move towards a more sustainable world in which we handle energy and raw materials differently, or to a system involving self-driving cars. This is what the transition management profession is all about:

a profession that will be handed new possibilities with the rise of Big Data. A very promising experiment involving the application of Big Data is the Mobile Territorial Lab (MTL)[17] in Trentino run by the MIT researcher Alex Pentland. There, he is attempting to learn more about human behavior and human interaction though mobile phone and other data. His professional area of social physics is building connections between the quantitative world of Big Data and the non-quantitative world of social sciences.[18] The learning experiences obtained here will make an enormous contribution to how we shape large-scale transitions in the future.

Chapter 6
Data Analytics Is the Society

Ethics in the new information society

How Technology Develops Autonomously

Deep Brain Stimulation (DBS)—where electrodes placed in the brain give electric signals—initially seemed very beneficial to a patient combating the physical symptoms of Parkinson's disease. However, his behavior changed significantly as well. The man started a relationship with a married woman and bought her a second home and a holiday home abroad. He also purchased multiple cars, was involved in car accidents and had his driver's license revoked. Remarkably, the man was not aware of the changes in his behavior until the DBS was switched off. However, switching off the DBS made his Parkinson's symptoms return so violently that he became completely bedridden and dependent. The sad thing about the case is that there did not seem to be any middle ground. He had to choose between a life as a bedridden person with the symptoms of Parkinson's and a life without inhibitions, which would get him into huge difficulties. In the end, he chose—during a period when the DBS was switched off—to admit himself to a psychiatric hospital, allowing him to continue his life with the DBS switched on and therefore fewer Parkinson's symptoms, but better protected from himself.[1]

Are humans capable of handling the rapid integration of technology in our lives? Or are we slowly destroying ourselves? It is an issue that has been much debated in recent history. The rise of the telephone decades ago resulted in complaints that the

© Atlantis Press and the author(s) 2016

S. Klous and N. Wielaard, *We are Big Data*,

DOI 10.2991/978-94-6239-183-3_6

demise of close social contact was imminent. Nowadays we know better, although it must be said that in 2016, many parents do not understand why their kids are continuously 'WhatsApping' and 'Instagramming' at social gatherings. These concerns are not so different from the concerns parents had half a century ago.

With the rise of the new information society, the questions regarding the effects of the technology seem to be getting more profound. Are we as a society ready to take our most important decisions based on data analyses? Can we still make independent choices and form opinions? Will we still be behind the steering wheels of our lives? And when algorithms start doing more and more for us, will we as a society not become disconnected from ourselves, as was the case with the Parkinson's patient that was not himself anymore when he was 'supported' by Deep Brain Stimulation?

New ethical dilemmas, which we as consumers have co-created, because they have everything to do with our extreme expectations, will develop in the new information society. Those expectations are driving the speed of innovations to the extent that Oxford University scientist Nick Bostrom wonders whether we as humanity are capable of coping with the speed of the developments required to meet our extreme expectations. "Before the prospect of an intelligence explosion, we humans are like small children playing with a bomb. Such is the mismatch between the power of our plaything and the immaturity of our conduct. Superintelligence is a challenge for which we are not ready now and will not be ready for a long time. We have little idea when the detonation will occur, though if we hold the device to our ear we can hear a faint ticking sound."[2] Bostrom is worried about the speed at which we are racing and asks himself whether humanity—which sees itself as the king of evolution—is capable of managing its own innovation urges. He foresees existential damage to humankind. We will have to develop a plan on how to deal with rapid technological developments in order to have a future as a society in which we feel comfortable.

This is not a simple problem and a straightforward answer is not easily found. However, in this chapter we will try to make a serious start.

Technology Has Two Faces

In this book, we already referred to the fact that technology by its nature has a light and a dark side. We believe that the challenge is to use technology such that on balance we make progress. This requires conscious decisions about how we deal with technology, and it is questionable whether we are capable of making those decisions. As we wrote earlier, people are herd animals and creatures of habit and therefore are only capable of making—or willing to make—their own conscious choices to a limited degree.

It is vital, especially with regard to data analyses, that we make choices about the measure in which we allow them to be used to make our lives more pleasant and under what conditions. Sometimes this is not a difficult choice. For example, take a car where the low-beam headlights are switched on automatically when a sensor signals that it is starting to get dark. That sensor is—compared to humans—not sloppy or forgetful and the advantages clearly outweigh the barely-existing disadvantages. Such tools are already deeply embedded in our lives, from a thermostat that decides what time to turn the furnace on in the morning based on weather conditions, to the onboard computer in our car that automatically dials emergency services in case of a serious accident. These tools are very helpful, because they allow us to leave many decisions to technology.

We Are Becoming Increasingly Dependent on Technology

The degree to which our decisions depend on the data that technology gives us will only increase. Data analysis will ensure that the self-driving car will take over many decisions from the driver.

This car will be equipped with a large number of sensors that generate data flows to let the car maneuver safely through traffic, maintain it properly, and make supplementary services possible. Your car will not only determine for you what the safest and fastest route is, but also when it is time for maintenance or an update of the systems. In theory, there will be no more traffic violations because we will be driven according to the parameters set by traffic legislation. However, what if an accident does occur? A judge may conceivably have to determine, based on data, which system has failed.

This scenario potentially poses life-sized dilemmas. Consider the following. When driving on a mountain road, a car meets a group of descending cyclists in a hairpin bend. It is too late to avoid the cyclists and the car has two options: drive into the ravine itself, with fatal consequences to the driver, or collide with the group of cyclists. In that case, data analysis even plays a role in an ethical dilemma. We are probably capable of making a good self-driving car, but can we also make a self-driving car with feelings? A car capable of making moral decisions?

The self-driving car is a symbol for the manner in which the information society is developing. The truth of the matter is that society has always been a data analytics ecosystem: because humanity has continuously intervened in our existence using data analysis. We can differentiate three stages in this respect:

1. **The tool stage.** In this case, the effort required is provided by society, and the intelligence is also provided by society. An example is the abacus.
2. **The machine stage.** The effort of the data analysis is made by a machine. However, society determines the objective for which the analysis is used. An example is the calculator.
3. **The automate stage.** Both the effort and determining the objective are done by data analysis. In a certain sense, this makes society dispensable. The data analysis operates autonomously. An example is the chess computer that not only

executes calculations, but also formulates strategies. In principle, two chess computers could meet their objectives face to face with each other, without any intervention from society. In this third phase, we are on our way to what is known as the singularity.

The moment machines are more capable of making decisions than humans, without us being able to understand why that decision is better is known as 'technical singularity' and is described by Vernor Vinge in his essay *Singularity* from 1993.[3] Furthermore, Vinge argues that progress is pushed by intelligence. If the intelligence of machines is growing, more progress will be made, which will lead to a higher intelligence level in the next generation of machines. He expects that our current models and rules will no longer work in this spiral. Therefore, he uses the word singularity, a term that points to a discontinuity in the behavior of a system in time. This singularity appears to be getting closer if we extrapolate current technological developments.

In this book, we will not discuss the world after the technical singularity has occurred. It is already complicated enough to keep the evolution into a new information society manageable up to the point of singularity. We are, therefore, significantly limiting ourselves. It means that in this book we assume that humans are still capable of understanding the analyses of computer systems. We would consider it an undesirable situation if this were not the case in the current pre-singular world. That means that the final responsibility for the correctness of a decision lies with humans, even though that decision was made based on a data analysis.

Note that decisions based on data analysis are already difficult for people to understand at the moment, in some cases. An example is the trade in complex derivatives that contributed to the start of the credit crisis. This trade was dominated by systems that calculated risks, without the users of the systems being able to understand them. This was, in our frame of mind, an undesirable situation.

The evolution of the role of data analysis in our society that we sketched—we estimate that we are between the second and third stages, between machine and automate—brings with it some large ethical issues. Increasingly often, images of a world in which humans are becoming the obedient slaves of technology are looming.

Is Data Analysis a Danger to Our Democracy?

Critical thinkers in this respect include Dave Eggers (author of the bestseller *The Circle*) and Evgeny Morozov (author of *The Net Delusion: The Dark Side of Internet Freedom*). The latter sees in our digital future a big danger to democracy, because our freedom is increasingly limited by what he calls "a border of invisible barbed wire".[4] This danger comes from the fact that decisions are increasingly made by algorithms, based on data we put in. The problem in this respect is that it is getting increasingly more difficult to explain why a decision was made, to see the reasons behind a decision and to see whether or not these reasons were valid. An example that rose from his own mind helps to explain this.

Suppose you want to buy a T-shirt and find that they are manufactured in Bangladesh. As a consumer you are then faced with a dilemma. If you decide to buy the T-shirt, you may be contributing to dangerous and unhealthy labor conditions there. Should you not buy the T-shirt you may cause a fall in employment in the textile industry in Bangladesh which may result in a child having to enter prostitution. In the information society, we look for answers to these dilemmas by performing research on the Internet. In fact, we put these considerations into the hands of a search engine that maybe—or probably—only shows us part of the truth.

According to that reasoning, technology is herding us with invisible barbed wire towards a world in which we hardly make any choices ourselves anymore. But really, there is no alternative. The result is that we no longer make independent evaluations, but leave it to the information provided to us by a search engine. That machine is, of course, dependent on our input, but is definitely not objective. What's more, it remembers what we have asked before in order to give answers that fit our profile. In one of his essays, Morozov explains his nightmare scenario in which people no longer make choices themselves:

"The invisible barbed wire of big data limits our lives to a space that might look quiet and enticing enough but is not of our own choosing and that we cannot rebuild or expand. The worst part is that we do not see it as such. Because we believe that we are free to go anywhere, the barbed wire remains invisible. What is worse is that there is no one to blame, certainly not Google, Dick Cheney, or the NSA. It's the result of many different logics and systems—of modern capitalism, of bureaucratic governance, of risk management—that get supercharged by the automation of information processing and by the depoliticization of politics."[5]

Will the dark side of Big Data really lead us into a frightening world? A world in which people according to Morozov "(...) can relax and enjoy themselves, only to be nudged, occasionally, whenever they are about to forget to buy broccoli." But this is nothing new. For years, we have been getting biased information, for example, through television. If, nowadays, we see the Internet industry as a threat to democratic values, this was also the case for years with Hollywood's film industry. According to that same reasoning, the danger began as far back as the invention of the printing press. In short, we have been living in a world full of behavioral influences for a long time, both inside and outside the digital world.

What is more important is that this doomsday scenario thinking does not offer any real solutions. The suggestions are mainly based on ideological wishes, are trying to politicize the debate on privacy, and will probably not lead anywhere. Privacy has been an important theme in politics and up to now has not resulted in satisfying answers, even in the 'post-Snowden' era. We also do not believe in calling for civil disobedience, in the way that Morozov calls for an information boycott. This would, of course, be an excellent solution if large groups of people were to consciously think about their behavior and draw the necessary conclusions. The reality, however, is that only a small minority will be willing to go that far and therefore this solution is doomed from the start. The masses will embrace (new) technology with hardly any criticism because it makes their lives easier, or at least gives them the feeling that it does. Simply said, the masses like their convenience. Less than 1 % of people actually read the terms and conditions of newly installed apps. We thoughtlessly click away cookie warnings of websites. In the aftermath of some privacy incident or the next tapping scandal, some people will consider their own behavior critically for a moment before going back to what they were doing before.

Data Analysis and Society

We believe that the issue is not where we draw the line for applying data analysis in society, but rather how best to shape that interconnection. Dutch professor of Philosophy of Technology Peter-Paul Verbeek offers valuable insights in this respect. On 15 October 2009, in his inaugural speech on the philosophy of humanity and technology,[6] he argued for a new vision of the relationship between humanity and technology, in part because he is dissatisfied with the current mindsets. Those mindsets, of bioconservatives on the one hand and transhumanists on the other, can be almost seamlessly projected onto the positions that Morozov—'the system is feeding you broccoli', and Mayer-Schönberger—'how the data explosion will answer all our questions'[7] take in the debate on our new information society.

In fact, both approaches ignore the relationship between data analysis and society, which in our opinion are connected with each other at the deepest level. Ethical judgements on how far data analysis in society may go (Mozorov) are not appropriate in this respect. The reasoning that data analysis is a tool to solve almost all our problems (Mayer-Schönberger) is not satisfactory either. Nowadays, data analysis is more than a tool. With the support of data analysis, we create intelligent environments, resulting in immersion: being submersed in an environment that reacts intelligently to presence and activity. As a result, data analysis becomes Big Data—in line with the definition in our introduction—and in doing so rethinks and reshapes society yet again. This could be called a fusion of analytics and humanity.

Big Data will then also play a fundamental mediating role in our experiences and activities. The classic philosophical thought is that society is sovereign—or superior—compared to data analysis. In this line of thinking, we use data analysis simply as a neutral tool to reach the objectives set by us autonomously.

Modern philosophy focuses more on better understanding the relationship between society and Big Data. The central idea in that context is that society and Big Data cannot be understood separately, but only in their mutual relationship. They are not only often actually interwoven, but cannot be understood without each other. Therefore, another approach is required in the discussions on the ethics of Big Data; an approach in which we consider the information society as a system into which Big Data is integrated.

A fascinating example of the manner in which technology has become a completely natural part of ourselves and society became visible in the project known as the Hole in the Wall project of professor Sugata Mitra of the Newcastle University in England. In 1999, he simply made a hole in a wall in a slum in New Delhi, India, in which he put a computer with an internet connection. He wanted to see how street children—who had never previously worked with a computer—interacted with it. To his joy, they were able to not only acquire computer skills within a short time, but also to teach themselves English and math. They achieved this without a teacher, and without the encouragement of a project manager or scientists. Since then, the project has been expanded to various countries.[8]

Therefore, we have to view Big Data as an integral part of society. A comparison with the rise of language is possible: Big Data and its usage are developing like language itself, even if we do nothing. And it is becoming part of ourselves and society; it is not just a tool. It may even offer us meaning.

This approach leads to a fundamentally different ethical question. Instead of wondering whether certain types of Big Data are admissible, we can focus on the question of how to embed it in society. The key question in this respect is what kind of society we want to be and how Big Data can help us to achieve that.

This question is not only a pleasant pastime for writers of books, but is also an important issue for companies such as Google. In January 2014, Google bought Deep Mind Technologies, a company specialized in artificial intelligence (AI). The company develops computer technology that can think and act like a human. Deep Mind allegedly set a firm precondition to the deal: Google had to set up an ethics committee to deal with this subject.[9] The committee monitors the manner in which Google is allowed to use the technology. This is relevant in an era in which big steps are being taken with AI, which is also made possible by the increased calculation power of computers that enable 'imitating' neural networks. Google scientists predict that within 10 years, it will be possible for computers to use their 'common sense'.[10] They will be able to flirt with us and make the movie *Her* (2013) a reality. Other big internet companies are also highly committed to AI. In China, Baidu has opened an AI lab. Microsoft uses AI for speech recognition and Facebook uses it for face recognition for photos.

Time to Embrace Big Data

In light of the above, it will not be a surprise to you that we are not going to sketch doomsday scenarios about how the new information society will change us into zombies without choices or free will. It is our opinion that we should not see the development of Big Data as an erosion of our existence and the free choices that we can make. Big Data and society will increasingly fuse together. Slowly but surely, we will feel at ease with the comfort that it brings us and will integrate it seamlessly into our behavior.

In this respect, professor Verbeek points out that some people feel more like themselves when taking Prozac, which he sees as evidence that current technological developments are putting a radical end to the idea that humans are autonomous and authentic subjects whose characteristics we should be able to understand. We can draw a parallel between that position and the manner in which Big Data influences our lives. For example, we receive a tailor-made advertisement for new car tires because 'the system' knows that we have driven 50,000 miles on the old ones; we have a virtual assistant that gives us a digital poke when it is time to leave for our appointment because it can see that there is a traffic jam on our route; a music app introduces us to music that we will probably like because it knows what we love from our profile; when we are looking for our friends at a festival, we can localize them with an app. And we are okay with all this.

In Chap. 5, we argued that Big Data may contribute to solving the sustainability crisis in which the world finds itself. Big Data makes it possible to develop new mechanisms that motivate or even force us to contribute to a good society. Or at least to what we have defined as a good society.

Many articles and opinion pieces put Big Data in the category of evil concepts that are threatening society, as if it is something that has to be restrained and restricted in order to safeguard our freedom. That is understandable because indiscriminately accepting it involves many dangers. We have to consider Big Data as an integral part of the information society however, so that we can have an ethical discussion about it. Not only about Big Data as such, but about a society *with* Big Data, a society that is continuously reinvented by new applications of Big Data. We can embrace that as a form of progress, as long as we maintain the basic rule that we understand what we are doing with Big Data and why we are doing it and retain the ability to choose ourselves what we find acceptable.

If we do that, a society that takes responsibility for its own existence will develop. A society that discusses fundamental frameworks and does not just deal with the fallout from incorrect analyses performed by, for example, banks or insurance companies. In order to achieve this, we require different competencies than those with which we are currently equipped. We have to start thinking differently about the profile of data scientists, but above all, about the entire education system in a wider sense.

Chapter 7
Wanted: Thousands of Sherlock Holmes Clones

Data analysis requires the switching off of preconceptions and expectations and the switching on of common sense

Spurious Correlations

In 1999, US government spending on science, space exploration and technology amounted to $18.1 billion. The number of suicides in that year was 5427. Ten years later these numbers were $29.4 billion and 9000 respectively. The increase was similar, relatively speaking and the correlation over the years was an incredibly high 99.8 %. A similar connection can be found between the number of movies in which Nicolas Cage appears and the number of people that drown by falling into swimming pools, although that correlation is admittedly slightly less convincing. Using such farcical examples, Tyler Vigen shows how we can fool ourselves with data on his 'spurious correlations' website.[1]

In previous chapters, we discussed how Big Data can contribute to societal changes and what ethical challenges we can expect to encounter along the way. It is essential that we are capable of translating the increasingly larger quantities of available data into meaningful decisions. This requires proper data analysis which does not always happen. We will address this point in this chapter.

The fictional detective Sherlock Holmes hits the nail on the head by stating: "the world is full of obvious things which nobody

© Atlantis Press and the author(s) 2016
S. Klous and N. Wielaard, *We are Big Data*,
DOI 10.2991/978-94-6239-183-3_7

by any chance ever observes" and "there is nothing more deceptive than an obvious fact."[2] These quotes constitute one of the key warnings that should be issued to everyone that wants to use Big Data. We have to prevent data analysis causing problems due to poor interpretation. If you collect enough data and perform enough different analyses, you will always find an interesting correlation. If you recklessly abuse the calculation power of a computer, you will come to many conclusions that are not very relevant or that are even misleading. One example is the supposed correlation between butter manufacturing in Bangladesh and the annual returns of the Standard & Poor's 500 stock index.[3]

Against that backdrop, there is increasingly more criticism that Big Data will only lead to more (social) accidents. The reasoning behind that criticism: being able to model large amounts of data is only meaningful if we do not forget the context of the data. Without that context, data loses its value. The previously mentioned flash crash of 6 May 2010 demonstrated that once more.[4]

At that time, computer programs, based on their algorithms, initiated a sell wave. In 13 minutes, hundreds of billions of dollars were wiped off stock values. Trading was suspended for five seconds, after which the stock market quickly recovered. The crash happened because of 'high frequency trading', where computer programs operating at a phenomenal speed buy and sell large bundles of shares based on the data fed to them, and where one millisecond means the difference between profit and loss. The decisions are made by computers and the underlying algorithms are often not even understood by the traders themselves. Therefore, the context is absent.

Use Data Wisely

One of the prominent critics of Big Data is Nassim Nicholas Taleb, who says "Big Data means Big Errors".[5] According to him, analysis of statistics will result in completely incorrect conclusions and the role of coincidence is underestimated. He is right. As contradictory as it may seem, this is exactly why more investment in the area of Big Data research is needed.

It is a fact that the more data you examine, the more patterns you discover that are purely coincidental and will not repeat themselves. Care is needed, especially in an era where people sometimes believe that Big Data is a machine into which you pour large volumes of data, wait for the machine to finish its number crunching, and receive ready-made solutions. The reality is often much more complicated. Successfully applying Big Data is often a process full of setbacks—a process involving blood, sweat, and tears.

In this respect, it is very important that we do not forget the difference between correlation and causation. In the case of correlation, there is a statistical connection between factors, but not necessarily a cause and effect relationship. Taleb gives the example of the statistical connection between the duration of a hospital admission and the zodiac sign of the patient. That is, of course, a correlation of zero value. But in the case of causality, there is a statistical connection *and* a connection that can be explained. For example a harsh winter results in higher energy bills. Then there is the middle ground. Let's call it 'the early warning'. In this case, there is no clear cause and effect relationship, but there are strong indications that there is a causal connection somewhere that connects the two together. Think about a boxer who (unknowingly) betrays himself through small movements to his opponent who can then anticipate where the next move will come from.

In some cases, it is not necessarily a bad thing if we do not know why there is a certain connection. Viktor Mayer-Schönberger uses the example of researchers at the University of Ontario, Canada, that collected large quantities of data—1000 data points per second—from premature babies to demonstrate this. One of the main risks for these incubator infants is that they may catch an infection and a rapid response by medical staff is of the utmost importance. The researchers were able to identify infection patterns and could predict that a baby was getting an infection 24 h in advance of any symptoms appearing. Strangely enough, the infants' vital signs became more stable in the period before an infection was confirmed, not more instable. Why that pattern occurs is not yet known, but how bad can it be if it saves the life of a child? The data points form an early warning system, which in this case is sufficient to prompt action.

The opposite may be true in other cases. We don't want a situation such as in the movie *Minority Report*, in which we preventively arrest people based on data patterns without understanding why these data patterns point in the direction of a planned murder.

This is, of course, far removed from the current social reality and therefore not a real danger yet. Other cases are getting closer to home, however. For instance, we know from our own experience that it is possible to predict years in advance using large quantities of data from various sources whether someone will get into financial difficulty, probably even before the person knows it. In a situation like that. Whether we should use data in this way is definitely not an open and shut case, if only because it could result in self-fulfilling prophecies. The real danger of a data-soaked society is that we draw solid conclusions to correlations. Suppose that the revolution in the health sector—as mentioned in a previous chapter—actually happens. We would hope that common sense is not switched off and that a sensible scientist looks critically at the results of the analyses. We prefer to be treated based on causality, not correlation.

Simpson's Paradox

What have we learned from this? We have to keep a sharp focus on making sure that data analysis is not executed lightly and that patterns are not translated into conclusions without cause. Sherlock Holmes stories offer a wealth of wisdom for everyone that works with data analysis. One example from an almost end-less number of quotes, "It is a capital mistake to theorize before you have all the evidence. It biases the judgement."[6]

A good data scientist is aware of the risks we mentioned and is trained to be very critical of them. Simpson's paradox, which can easily be explained to non-statisticians with an anecdote, is central to this. Statistics show that sailors that go overboard without life jackets are saved more often than sailors that wear a lifejacket. That goes against intuition, but on closer analysis it can easily be explained. Sailors mostly opt to wear life jackets in bad weather conditions—conditions in which a rescue is difficult or even impossible. However, this example should not lead to the conclusion that wearing life jackets is a bad idea. This example shows how important the context of data is to drawing responsi-ble conclusions.

This example is typical of the world in which a data scientist operates: when you combine data in a clever manner, unlikely results are sometimes obtained. An incorrect conclusion can have fatal consequences. This would not be the first time that the deci-sion has been made to stop using life jackets based on data analy-sis—figuratively speaking.

We have to prevent these accidents now that Big Data is becoming increasingly interconnected with society. Therefore, we should not leave data analysis entirely to a powerful computer that can do very clever things. Number crunching is only the (rel-atively easy) start of an analysis, the quality of which depends on the competencies of the data scientist. The difficulty lies espe-cially in being able to understand or interpret the results, possibly resulting in interesting insights for the person commissioning the study. We should properly train data scientists that are capable of dealing with Simpson's paradox so that they are sharp and criti-cal when interpreting data. This will allow them to search for true causal relationships without any tunnel vision, which they can go on to interpret in collaboration with other experts.

Good data scientists are not a luxury because data will play an increasingly important role in society and we very much need these data scientists in order to prevent accidents happening with data. Especially in an environment where we have access to large quantities of data—big and messy—it is often not easy to understand relationships between variables as in the relatively simple case of the lifejackets. For instance, in Chap. 1, we already showed how the measurements of the particle accelerator in CERN were influenced by the position of the moon. This emphasizes the necessity for data scientists to be more than clever boys and girls that are good with statistics. They have to be capable of working like a true Sherlock Holmes. They have to be able to deal with setbacks when researching and understanding significant results. Moreover, they have to be able to understand the context from which the data was sourced like no one else and like never before. Because of this, groundbreaking Big Data projects will never be a plain vanilla affair.

Limits to Human Understanding

In this respect, however, we will reach limits. It is inevitable that there will be more and more cases that even the best data scientist will not understand. This is already happening in some areas. One example is the financial sector where we can still question whether people understand it. For example Deutsche Bank in Germany had an astronomical outstanding amount of €54 trillion in financial instruments in 2013.[7] This amount cannot be found anywhere on the bank's balance sheet because from an accounting perspective, it is possible to offset certain instruments against others, but that is not relevant here. The scope of this portfolio is not only larger than that of every bank in the world; it is also more than the sum of all global economies combined. It's difficult to explain that concept to the average consumer. Even seasoned professionals don't understand it completely. Stefan Krause, who was the CFO of Deutsche Bank at the time, exemplified this when he said that the bank's accountants were not actually capable of calculating how much capital they should keep as a stable buffer for the outstanding obligations.[8]

It seems inevitable that in the future we will increasingly be incapable of interpreting our own world. As former Google CEO Eric Schmidt once put it, "The Internet is the first thing that humanity has built that humanity doesn't understand, the largest experiment in anarchy we've ever had."[9]

Chapter 8
The Question Is More Important Than the Answer

Educational institutions have to be given another role

Socrates Reinvented

In the traditional passive education model, teachers are the owners of knowledge and information; they ask all the questions. Students are expected only to absorb the information they will later need to reproduce on tests. In the book Make just one change Dan Rothstein and Luz Santana argue that it should be the other way around: students should be taught to ask their own questions. In fact, their theory could be seen as a reinvention of an old philosophical concept, that of Socrates wandering around Athens asking questions to get at a deeper truth.[1]

When we extrapolate the previous chapter and see Big Data as an integral part of our society, we cannot ignore the fact that we have to increase our knowledge, know-how and awareness of this subject. We have to become better at asking critical questions with regard to its applications in our daily life. This does not apply only to a specialist data scientist, who should have Sherlock Holmes-like critical questioning skills. It applies to everyone who allows this type of analysis into his or her daily life—and that means (almost) all of us.

We have to understand where information comes from, on which sources it is based and how it came about. We have to be able to determine how reliable certain information is; we have to learn to assess what information is relevant to make a decision;

© Atlantis Press and the author(s) 2016
S. Klous and N. Wielaard, *We are Big Data*,
DOI 10.2991/978-94-6239-183-3_8

we have to learn to understand which decisions are important and which are not. We also have to learn how that impacts the level of quality required in the information we use to arrive at that decision.

In short, we have to learn to ask questions. In this chapter, we discuss why that is essential and which changes are required.

The Reliability of Information

Let us first analyze how we obtain our information. The Internet is a jungle of reliable and unreliable information and it is quite a job to find the right information. Since the world wide web is already decades old, most users understand that we have to be critical with regard to the reliability of the information we find there. A blogger that is promoting a new camera with a bit too much enthusiasm is probably doing so because the camera's manufacturer is paying him. With every news report we see, we have to ask ourselves who is behind that news and how certain interests—commercial, political or even personal—could have colored it.

So far, this is nothing new. The risk of unreliable or colored information has always existed. It is not for nothing that an old cynical journalism joke attributed to Mark Twain says that you should never let facts get in the way of a good story. What is new is that the plurality in the current information society has increased because anyone can become a journalist and publisher. Following a reliable information diet has become much less easy due to the multitude of online information sources from which you can choose. At the same time, the risk of blunders has increased, because the shortened news cycle means that news media are under a lot of pressure to act fast. Furthermore, the Internet has the characteristic of an open microphone combined with an infinite memory; once a story gets out, it keeps resonating and turning up in search engine results. Fighting against untruths is often impossible. The story has already been framed in our minds and it will stick there. Therefore, the impact of an error can be big. Once Bloomberg repeated in error an old story about financial problems at United Airlines. The message was picked up as current news and the United Airlines stock price took a significant hit.

Both obscure bloggers and renowned news media can get it spectacularly wrong. In 2003 at *The New York Times*—often considered as one of the world's best newspapers—the chief editor had to step down once it became known that one of his star reporters had made up dozens of stories.[2] The credibility of CNN received a heavy blow when the news channel had to withdraw a story about the use of poisonous gas. A movie called *Shattered Glass* was made about the journalist Stephen Glass, who invented dozens of stories that were published as fact in the renowned opinion magazine *The New Republic*. Some of his stories were based on Internet 'facts' that he constructed.

These examples make it clear that it is essential to remain critical of the truthfulness of the information presented, whatever the source. Following a diverse news offering should be one of our basic principles. It is questionable whether we are sufficiently aware of the necessity of adopting a critical attitude with regard to the information we find on the Internet since we are now almost continuously connected and convenience seems to have become our key motivation. Opinions are divided in this respect. Some believe that the young generation that is being brought up in this new information society, is more capable of finding its way through the information jungle; young people know that world of information switches between hard facts and popular opinions and they surf together their own truth. Others fear that the urge for convenience results in mindlessly accepting information.

You Cannot See Everything Through
Your Own Small Window

As well as the risk of colored or incorrect information, there is another question: are we seeing all the relevant information or are certain issues hidden from us? The latter is almost inevitable and is becoming increasingly more of an issue as the information we see becomes more personalized. Internet companies such as Google and Facebook provide us with tailor-made information. They increasingly know, as we discussed earlier, where we are, what searches we do and what we click on. They use that information to build profiles for each of us, in order to make news-feeds and search results more relevant to us as users. That is convenient—and many users would have it no other way—but it also has a side effect that has become known in the online world as the filter bubble. The Internet activist Eli Pariser wrote a book[3] about its dangers. When two friends do a Google search for 'Egypt', one set of search results mainly shows tourist destinations while the other receives information about the political crisis in Egypt. Personalized information is convenient, but it also creates blind spots because a lot information passes you by. It may be an annoyance when we cannot find what we are looking for, but it will become a bigger problem, for example, if a couple of people take out their smartphones at the same time and search together for 'climate change'. A person that often visits conspiracy websites to do research for his job will learn from his smartphone that climate change is a myth, whereas a scientist will be shown sites where climate change is presented as a fact.

The search engine that presents these various search results is doing so based on our profile and our behavior online in order to comply with our demand for personalized information. This is convenient because now we do not have to sort through all kinds of search results that we do not find relevant. But it also makes it very difficult to see the world beyond that. If you live in the hipster part of San Francisco, it is easy to start believing that the whole world resembles your neighborhood. The danger is that we create a cocoon around our digital life, making it difficult to look outside of that cocoon. In other words, the 'random' factor in information is slowly disappearing.

Today, it is still possible to get the information you want by trying a little bit harder. If you look beyond the first page of the Google search results, you will probably get less limited information. Now, even that has come under pressure, partly because of the discussion around 'the right to be forgotten'. This principle was part of a legal case in 2014 at the European Court. A Spanish lawyer, Mario Costeja González, demanded that Google would no longer return certain search results in response to his name. The information referred to a conflict that the lawyer had in 1998 with the Spanish tax authorities, which was reported in local newspapers at the time. However, even after the issue was resolved, the news kept popping up every time the lawyer's name was typed into Google. González held the opinion that this was a breach of his privacy and successfully demanded that the information be removed. The judge allowed privacy—'the right to be forgotten'—to prevail over freedom of expression.

The decision put Google in a different role. The company now has to act as a librarian that is obliged to close part of the library to its customers. We can think of no better example of a conflict between privacy and convenience. As an Internet user, you want convenience and, therefore, access to all of the available information. That is not the case for González, however, who wants his privacy.

It is unclear how far the impact of the judgement will reach and what the impact will be on (the operations of) Google and other Internet services. Wikipedia is also affected and its founder Jimmy Wales is not happy about being forced to remove certain information as a result of someone invoking the right to be forgotten. During a conference, Wales said that he thought it was "deeply immoral" and that it will result in an Internet "riddled with memory holes."

Since anyone can submit a request to remove unwanted results from the Google index or other sources, this has the potential to result in unworkable situations. As of October 2015, Google had already received 330,000 such requests.[4]

In any case, the conclusion is clear: search engines such as Google only offer a small window into the world. At first sight, that may just seem to be inconvenient, but it also comes with a fundamental danger. The filter bubble may lead to constant self-confirmation and even endanger democracy, to which collisions between various ideologies are essential. The more a person uses the Internet, the more personalized their window becomes and the less they are exposed to opposing ideas. The possible result is the radicalization of the person's own ideas, even with the ultimate result of extremism and violence.

Compartmentalization

It would be taking a narrow view to suggest that only search engines and social media drive the filter bubble. In fact, our own behavior on the Internet leads to compartmentalization. The web is home to numerous communities in which only kindred spirits are welcome and where hardly any real discussion takes place.

Looking Beyond Your Filter Bubble

In the new information society, we have to find the right balance between more convenience and an enormously wide offering of information. History teaches us that the masses will choose convenience. How many people are bothered to look for the 'other side' of the mainstream news shows? That is a rhetorical question. At the same time, there is also a wonderful side to the rise of the Internet as a news medium. Indeed, in the television era, it was almost impossible for people to look beyond their own side's reporting in times of war. It was not possible to access information from the enemy's point of view. Nowadays, the information we receive from our own news media is still as colored as ever—because it is seen through the prism of our own culture and prevailing opinion—but at least an alternative is available to us. An example is the crisis that started in 2014 in Ukraine. We know that the Russian view is very different to the Western view and if we look a bit further, we can learn more about this different view.

Nevertheless, the risk of an information ghetto is real since Internet companies want to keep improving their personalized services based on personal information. They will find new ways to retrieve exactly the information that is relevant for the profile of the user and that is considered meaningful by its peers—because that is probably the most relevant information for the user. A quote from the founder of Facebook, Mark Zuckerberg, says it all. He knows more than anyone that Facebook has to provide its users with relevant information and he once said in this respect, "A squirrel dying in front of your house may be more relevant to your interests right now than people dying in Africa."[5] The result? Facebook encourages us to look mainly into our own front yards and much less at the big themes that really matter.

Can technology solve the very problems that technology causes? This is almost a philosophical question and now science has become interested in the filter bubble. Also Internet companies view it as a serious problem. A collaboration between the Pompeu Fabra University in Barcelona, Spain and Yahoo Labs resulted in an algorithm that may be the antidote to the filter bubble.[6] The core of the algorithm is that the search engine not only shows news based on what peers read, but also what dissenting people with similar interests in other areas read. The researchers have tested the algorithm in Chile, where in November 2014, heated debates were held about the strict abortion laws in the country. The result was encouraging: people seemed to be open to having a meaningful debate when they were confronted with the opposing opinions of comparable individuals.

However, we should focus on not only a technological solution, but on increasing our own critical abilities. In the new information society—in which putting together a proper information diet is not straightforward and in which Big Data is interwoven with almost everything that we do—we cannot function properly without strong, independent critical abilities. We have to keep taking responsibility for our decisions, even if they are data driven. Our education should not be a mold that shapes us into end products with predefined characteristics, but should inspire us to ask questions. We have to be capable of judging for ourselves, be skillful in debate and able to analyze and argue our position well, and we have to be able to look at the world from the perspective of other people. This is not a small challenge and in the remainder of this chapter we discuss how our current educational system can respond to these demands.

Learn How to Function in the New Information Society

First, let's go back to the rise of Big Data in society. We said in Chap. 6 that data analysis has always been an integral part of society. Think about the previously mentioned example of the sensor that, after an entirely automatic analysis of its environment, 'decides' to switch on the dim lights of our car. What is new is that the impact of the analyses is getting bigger, we are finding more connections to be made, and we are becoming increasingly dependent on them. We increasingly base our decisions on analyses performed by systems, such as a simple search on the Internet. Therefore, we have to remain alert to ensuring that we understand those decisions and that we know what information they were based on.

If we fail to do so, things could go badly. For example, our navigation systems could take us the long way around to our destinations in order to increase the revenue of certain gas stations without us ever knowing that we have not taken the shortest route. There is an even worse scenario. Our voting behavior could be manipulated by personalized news and opinions based on our profiles. We may be formally free to choose what box to tick, but we are being manipulated by 'the system'. In a sense, we could say that our digital profile is voting for us. So how do we adjust our education system such that we are better prepared for this new information society?

It is quite unacceptable that the current education system in many countries is out of touch with the needs related to the new information society. In an essay on higher education, professor Dirk van Damme, who heads the Centre for Educational Research and Innovation (CERI) at the Organisation for Economic Co-operation and Development's Directorate for Education and Skills—writes, "Education educates young people for an economic reality that is not really there anymore, let alone that it prepares new generations for the reality of tomorrow." Furthermore, he questions whether we prepare ourselves properly for the "competence switch to non-routine skills e.g. by teaching people to make better decisions in the context of large scale unpredictability and uncertainty."[7]

It seems inevitable that both the form and the content of the education system need to be drastically altered, from primary through to higher education and across all subjects. In a nutshell, it is important that people learn how to function properly as individuals within the new information society. Higher education will need to cater for more specialized, targeted needs, including for vocational training to prepare us for changes to the nature of our work. For a civil servant, for example, it is relevant that he should learn to ask how Big Data could contribute responsibly to better policies. For an IT developer, it is important that he/she learns to develop data analyses that meet the demands of society (and those demands are not always technological). A general medical practitioner has to ask how Big Data can help his/her patients and what the (ethical) concerns are. What are the correct questions to ask the patient, but also, what are the right questions to put to the system that will help the doctor to make a diagnosis (the decision support system)?

That change is required not only in higher education, but also earlier. Children in elementary school still learn about the fall of the Berlin Wall. It is much more relevant, however, that they should know how to search properly, and how to interpret the answers they find. Another key point is understanding the context in which information is presented. Does a child from the former Soviet Union see the same information or does he or she— whether due to the previously mentioned filter bubble or not— see a totally different version of the same events? Children have to learn to understand how that is possible and how it reflects on their relationship with and understanding of an Eastern European child. This is becoming a more immediate need than ever before since children today often have direct contract with children from other cultures and countries through games and social media.

In a certain sense, asking questions has become the only valuable source in the information society. Answers are everywhere and can be freely obtained through a search engine, and we increasingly let computers supply the answers. However, asking questions is still the reserve of people. Therefore, within education we should not only be focusing on what is correct and incorrect information, but more so on developing critical capabilities.

That competence plays an essential part in making responsible decisions. We should develop it so that we do not become mindless slaves to the analyses that machines perform for us and so that we continue to making conscious choices about how data analyses can and may make our lives more comfortable.

This brings us to the American philosopher Martha C. Nussbaum and her 2011 pamphlet *Not for Profit: Why Democracy Needs the Humanities*. In this pamphlet, she used the education system in India as an example to support her argument that education should mainly contribute to shaping a democratic global citizen. That sounds logical, but in many countries it is still very much a distant dream. Globally, education is increasingly focused on driving more economic growth. This is the case in the United States, where education is mainly designed to facilitate the increase of individual income and the economic growth of the country as a whole. It is also true in Europe where according to Nussbaum, lecturers have never learned how to teach their students. The most important thing for them is how many times they are published in science magazines. Actual interaction between teachers and their students is rare.

The somber image of reality sketched by Nussbaum is far removed from the classic academic ideal of Bildung.[8]

Fortunately, there are various signals that change is coming to all levels of education, in terms of both form and content. There is an acknowledged dissatisfaction about the educational system in the Western world and alternatives are being explored and initiated. These include modular courses, with more emphasis on themes and competences instead of the classic approach based on subjects and knowledge.

One example is known as the Open Educational Resources movement, where materials are made freely available for use and sharing. This phenomenon led to the development of Massive Online Open Courses (MOOCS) which provide online access to courses taught at high-quality educational institutions such as Harvard, Stanford and MIT to almost everyone in the world. Technology makes new forms of education and knowledge distribution possible. Therefore, education is also competing across geographical borders much more than before. Educational institutions keep each other sharp and force change. There is—at least in theory—a democratization in the access to education. Anyone with a good idea for a different approach can start a course and may attract a large audience. We do not have to constrain ourselves to complying with the rules of 'the system' and we have the power to do things differently, in the same way that we as a collective of people can force governments and companies to change, as we described earlier.

Interestingly, it is less easy to force change in education. Even in the open education system, the most popular courses are those offered by prestigious, often old-fashioned institutions. Why is this? We probably select the safe, familiar route, knowing that a degree from Stanford looks better on our resume than one from a new college that may offer great things but is completely unknown. In doing so, we are therefore frustrating the very change that we ourselves want.

Changing Form and Content

It is also difficult to change the content of education, partly because course content is regulated according to national and international quality standards. The curriculum, therefore, only changes slowly, especially in primary education. Reacting to new (technological) developments is, by definition, a difficult challenge. In the case of the rise of the information society, it is perhaps even more difficult because its influence is not just limited to some subjects, but extends across all curricula and all subjects.

It is a bit easier to change the form of education, even though there is a lot of resistance to this. Many parents base their choice of school on tradition (the past) rather than innovation (the future) and they mainly consider the reputation of a school. That reputation is often related to the number of applicants that are refused entry, rather than whether the education it offers is capable of influencing the mindset of its students and stimulating creativity. Therefore, concerned parents often stand in the way of innovation, based on the fear that their child is being used as a guinea pig in a non-traditional environment.

Time for Change

"In times of change learners inherit the earth; while the learned find themselves beautifully equipped to deal with a world that no longer exists."[9] This quote from American author Eric Hoffer makes it perfectly clear that education needs to change. Educational institutions have to educate their students for the world of tomorrow. The dilemma is that the most prestigious educational institutions are proud of their long tradition of providing top-quality education. It is probably that very tradition that is hindering the speed of social change.

In simple terms, education is a game in which students can win by following the winning strategy. That strategy is about knowing the 'right answers'. The person who knows the right answer wins the game. Scores are kept in a report card. This game is only possible when the relevant information is segmented into boxes we call 'subjects'. Within those subjects, there is no doubt and, therefore, little room for questions.

That game is not, however, a reflection of the real world. The real world is not tightly segmented, and in our daily lives we are faced with all kinds of difficult questions that were not discussed during our courses. It is clear that neatly sorting problems into boxes does not (any longer) answer our needs.

We can look at knowledge as a network in which each small fact is a node. In that network, knowledge does not only increase as more nodes are added, but also and even faster when the number of connections and the strength of the connections between the nodes is increased. It is mainly through those connections that knowledge acquires its value. By only offering knowledge in neatly pre-shaped bundles, we do not properly prepare students for reality. While many educational institutions have begun to realize this, it has not so far precipitated any significant changes in the form and content of education.

In this context, the Dutch Minister of Education Jet Bussemaker mentioned "astronaut skills."[10] She referred to the problems Apollo 13 encountered when one of its oxygen tanks exploded in space. The solution to the emergency was found by using the lunar module and not the command module. Not only did the crew have to use cardboard and duct tape to fix the filter systems, but also they were faced with a greater challenge: how to return to earth. Without a navigation computer, they had to find the correct orbit around the planet based on the position of the stars and earth.

When they got back into orbit, they had to crawl into the damaged command module because the lunar module was not designed for a return into the atmosphere. As we know, the astronauts were successful in achieving this amazing feat. As Bussemaker said, it was "[...] a situation that not only required everything from their knowledge, but also their skills, It required everything from their creativity, their collaboration, their communication. It required everything from their courage, stamina, and flexibility. It required everything from them as people."

What we need is an educational approach in which we learn to make connections. An approach that stimulates flexibility in our thought patterns, based on information that is currently available to us. An approach that awakens creativity in order to combine knowledge in a new way, in order to solve the questions that are part of the new information society, in which everything is connected to everything else.

In such an approach, we focus on [11]:

Connecting students with other students and the ideas of others because valuable knowledge and ideas come into being between people and not in isolation.

Making connections with people outside of the institution, resulting in an ecosystem with regular and efficient connections with 'smart' external people.

Stimulating communication, collaboration and sharing of knowledge.

Building the skills needed to assess the value of information. Large quantities of information are freely available, but require the user to be capable of distinguishing signals from noise, to assess its quality and to improve the efficiency and effectiveness of the search process.

Stimulating active rather than passive learning. This requires an environment in which people can learn to do things instead of only reading about them, such as prototyping or practical assignments.

Creating more challenges by using games, competitions and new media. Learning can be very inspiring, but educational institutions often underexploit the possibilities.

These issues almost exclusively concern the form of the education and hardly touch its content. We already concluded that change can be achieved more easily through the form than the content and we see this development happening already. It should be achievable to execute this change, which will in turn dramatically change the role of the educational institution. It will no longer be about building knowledge, because knowledge is freely and easily available. It will be about advising and training students in order for them to develop a mindset with which they can face the challenges of the new information society. The educational institutions—and the teachers—will then no longer be the authority on which students depend for new knowledge. They will instead be advisors that prepare students for the new and rapidly changing world, a world in which they can ask the right questions and demonstrate creativity when it is needed.

Chapter 9
Transparency, a Weapon for Citizens

People themselves force change and do not need legislation to do it for them

Signposts for Air Traffic

The emergence of commercial airliners in the second half of the 20th century was a big social change. In the pioneering years, proposals were made to control increasing air traffic by painting signposts for planes on rooftops.[1] It seemed logical at the time, based on the idea that planes fly low and that they would be owned and flown by individuals. Today, we know how far that idea was from reality.

Woodrow Wilson, the 28th president of the United States wrote in 1884 that "Light is the only thing that can sweeten our political atmosphere [...] light that will open to view the innermost chambers of government." Fast forward to 2014, when the whistleblower organization WikiLeaks stated something similar on its website: "Publishing improves transparency, and this transparency creates a better society for all people. Better scrutiny leads to reduced corruption and stronger democracies in all society's institutions, including government, corporations, and other organizations."[2] This immediately raises questions: what can we as (a collective of) citizens do with the enormous amount of data to which we have access in the new information society? Will it make us more powerful and can we use it to keep companies focused? Should we put limits on transparency or make all data

© Atlantis Press and the author(s) 2016
S. Klous and N. Wielaard, *We are Big Data*,
DOI 10.2991/978-94-6239-183-3_9

public and accessible? What legislation is required to control the new information society? These questions will be discussed in this chapter.

Outdated Legislation

Let's start by stating that legislation almost always lags behind social developments. That applies in the new information society as much as anywhere else. An example is the simple question of whether the contents of an e-mail should have the same legal protection as a letter. Of course, it sounds completely logical that from now on, e-mails should have the same status since they have almost completely taken over the function of letters. However, in many countries there were decades of increasingly intensive email traffic before legislation was even drafted. This is emblematic of laws and legislation where changes have historically trailed behind social changes and now simply cannot keep up with changes in technology.

The question whether legislation and supervision can solve the problems mentioned in earlier chapters is answered by this example. For years now, the relationship between legislation and the Internet has been shown to be difficult on various fronts.

In Chap. 10, we will show that there still is potential for laws and legislation to influence the new information society, but not in the same way as they currently do. The previous examples do show, however, that we should not have too many expectations about the extent to which laws and regulations can help us to minimize the risks that are inherent to technology.

In fact it is questionable whether it is worrying that legislation is lagging behind since technological innovation gives people new possibilities to intervene that are possibly much stronger and faster. Individuals have much more influence than ever before because the Internet offers them the possibility to unite quickly in large and powerful collectives. There have been many cases in recent years that have proven that 'the people' can draw a lot of strength from the Internet in order to get issues on to the public agenda. One example is the Kony 2012 viral campaign.[3] This documentary by an activist who wanted to see the Ugandan rebel leader arrested spread like wildfire in 2012 and has been watched over 100 million times on YouTube. Its approach was much criticized. However, the consensus remains that it contributed to the decision of the African Union to send troops into Uganda to pursue Kony.

The growing influence of citizens is fed by two factors.

In a radically transparent world (almost) all information is available

Journalist Stewart Brand, who founded the Whole Earth Catalog in the 1960s, already stated "information wants to be free."[4] Information can be kept private for a long time, but once it is public, it will never be private again. The genie is completely out of the bottle, it is a one way arrow. In a sense, information follows the second law of thermodynamics: the result is maximum entropy. The endgame is chaos in which no data is stored in a safe container any more, but in which all information circulates in the public, uncontrolled space. That will never again change because you can't unscramble eggs.

The evolution toward maximum entropy is therefore inevitable; data that is freely available can never again become not freely available. This applies not only to personal data, but also to companies' and governments' information. It presents an entirely new communications challenge for organizations. The traditional view on managing a company's reputation—irrespective of whether it was for the benefit of the capital markets—was that the least amount of unfavorable information should be made public. Bad news should be camouflaged, good news emphasized.

The increasing transparency shows, however, just how meaningless that strategy is: bad news will get out there anyway. American publicist Clive Thompson introduced the concept of the see-through CEO and presented an interesting paradox:

> "The reputation economy creates an incentive to be more open, not less. Since Internet commentary is inescapable, the only way to influence it is to be part of it. Being transparent, opening up, posting interesting material frequently and often is the only way to amass positive links to yourself and thus to directly influence your Googleable reputation. Putting out more evasion or PR puffery won't work, because people will either ignore it and not link to it—or worse, pick the spin apart and enshrine those criticisms high on your Google list of life."[5]

In the DIY society, organizing is a piece of cake

Gradually, a DIY society is developing in which we as citizens do not or hardly need institutions and as individuals we are becoming more powerful than ever before. We should have seen this do-it-yourself society coming a long time ago. In 2001, two Internet friends in San Diego started an experiment called Wikipedia. The idea was too ridiculous for words: have thousands of people spread across the entire planet maintain an online encyclopedia, do not pay them a dime and do not have a management layer. Instead, make sure that every volunteer is allowed to amend or question the contributions of others. You would expect that such an anarchistic collaboration structure would result in chaos and inferior quality, but the opposite is true. Because all of the changes are reviewed very critically by other volunteers, there is good quality control. In the intervening period, we have learned that this model works fairly well—although, admittedly, not without errors.

Wikipedia teaches us that people are very capable of finding each other and collaborating without being managed. What's more, if they want they can easily bypass institutions—in this case the famous Encyclopaedia Britannica. The Internet is therefore proving to be a great tool for self-organization and it stimulates people to do things themselves. Not necessarily as individuals, but with each other.

That self-organization—in combination with the undermining of old institutions—is rapidly gaining ground. Are there more examples? Locally organized charities are becoming serious competitors of big charity organizations that have lost the trust of the public; wealthy people are bypassing banks and funding entrepreneurs with good ideas on a growing number of crowd-funding platforms. In many cases, the connecting power of social media plays an essential role. This also applies to Airbnb, a plat-form where people connect with others to rent rooms or houses. In the travel sector, this is now seen as a very successful form of self-organization. Something similar is happening with Uber, the fast-grwoing ride-hailing taxi service: a smart review system com-bined with social media is creating the trust required for stran-gers to transact and get into cars with each other.

This self-organization often also makes it possible to thumb a nose at the establishment, including politicians. It has been years already since the British newspaper *The Guardian* decided to put all expense statements of politicians on its website after a scan-dal regarding the expenses of British politicians. The objective of the newspaper was that the public would analyze these expenses statements themselves. It was also partly because the paper did not have sufficient capacity itself for such a huge operation. The collective was able to provide the labor required to scan the dis-organized data.[6]

We should also mention here that in 2011, some dictatorial regimes collapsed in part due to the use of social media. The pop-ular uprisings in the Arab world—including Egypt and Tunisia—were even called 'Twitter revolutions', because young people used Twitter and Facebook to communicate and organize upris-ings. Modern technology was one of the reasons that powerful dictators lost control of the people.

The Furrier the Problem, the Sooner It Is Solved

Based on the two previously mentioned factors—the DIY society and radical transparency—you could conclude that we are floating on a pink cloud and believe that the collective will solve all our problems as long as the world is transparent enough. That Twitter is stronger than all the bullets of dictatorial regimes. And that social media forces companies to only do good.

That is certainly not the case. Yes, social media played a big part in (the news on) the uprising in Egypt and other countries. Even more importantly, social media made an effort—despite all obstacles—to keep making that possible. After the regime blocked access to the Internet, Twitter and Google offered people in Egypt the possibility to send tweets to the world via their voicemail and YouTube opened a special channel for all videos from Egypt. Mainstream media gladly used all the material made available by these methods.

It must be acknowledged, however, that the results of the uprisings did not live up to the expectations most of us had when they took place. Furthermore, later analyses showed that the large majority of people on the street did not use social media and often did not even know what Twitter was. While social media did not start revolutions, it certainly strengthened them. If the seed of resistance is already there—or a general feeling of discontent—social media can feed it.

Furthermore, we are perhaps less idealistic in reality than we think. In 2013, over 1100 people died because a complex of small clothing factories collapsed in Bangladesh. The factories supplied clothing chains in the West that received sharp criticism because they contributed to the unhealthy and unsafe working conditions by squeezing their Bangladeshi suppliers. Over the years, various initiatives have been launched to do something about this, but the visible results are limited. Now that the disaster in Bangladesh is no longer at the forefront of our collective memory, any media reports that mention the issue point out that no fundamental change has taken place.

This is in stark contrast to the rapid actions that were organized after a shocking video appeared on how rabbits are treated to obtain angora wool. Fashion chain H&M decided within a few days to suspend the manufacturing of angora wool clothing and consumers were allowed to return angora wool purchases to the store. It appears that a cute, furry problem is sometimes taken more seriously than inhuman working conditions.

The New Creed: Undress for Success

Although there are, of course, limits to what a collective of people can and will achieve, a big change is definitely going on in how such collectives can force change in a transparent world. This means that businesses should move towards a more sustainable type of capitalism in which companies see themselves explicitly as part of society (and not separate from it), in which the focus is on the long term (and not on quarterly profits) and in which the needs of society weigh at least as much as those of the shareholders. This is where companies create the shared value as described by Michael Porter that we mentioned earlier.

This kind of transformation is a joint effort by companies, governments, and people. We believe that the key driver of the change will not be laws and legislation enforced by supervisors. The transformation will continue to be stimulated mainly by the extreme expectations of people, who will mobilize and simply force change as a result of the new far-reaching transparency.

In fact, businesses have no choice. They had better be radically transparent about what they do and why they do it, when the news is good, but also—and especially—when it is not so good. This will be the only way to maintain credibility and a moral license to operate. A lie has no legs, the truth will outrun it and in the new information society, truth runs faster than ever. The new creed for companies is therefore: undress for success. Show who you are without any marketing brouhaha. It is the only way to keep and maintain the trust of your customers.

The same applies to governments. They also need to let citizens peek behind the curtains. Big changes are happening in that area. When Barack Obama took office for his first term as president of the US, one of his election promises was more government transparency. In subsequent years, it proved to be one of the domains in which he was able to keep his word and took action. Almost immediately after taking office, Obama appointed a special senior civil servant, Vivek Kundra, whose only task as the 'CIO of the government' was to make all non-classified government information accessible to all Americans, freely and easily, entirely in line with the Freedom of Information Act. This law designates all government information as public property, based on the philosophy that taxpayers have already paid for it.

Disclosing this data through an Internet portal (data.gov) of course allows everyone to freely search an enormous amount of government data or develop applications for it. Many commercial parties were interested in developing such applications using data on, for example, traffic flows, government budgets or office buildings.

This phenomenon is called Open Data, a theme in which the United States is considered a trendsetter, followed closely by the United Kingdom. Other countries are also increasingly recognizing the importance of Open Data. In recent years, the European Commission has put it high on the agenda and requested member states to have implemented the Open Data legislation by 1 January 2015. The economic potential of Open Data is often referred to in this context. Then-European commissioner Neelie Kroes—a self-proclaimed fan of Open Data—believes that further disclosure in Europe is a true gold mine. According to her, at the time the legislation was passed in 2011, the economic value was already €30 billion and with the application of proper policies, it could be expected to increase to €70 billion.[7] The possibilities for creative entrepreneurs continue to be very wide-ranging and vary from better weather predictions to apps that accurately map which museum is or is not accessible to disabled persons.

There are many (technological) possibilities and the list of examples of applications is growing. One of the great applications of Open Data in the US is checkbooknyc.com, where New York City residents or other stakeholders can follow, at a very detailed level and almost in real time, the city's spending. The website even mentions whom the subcontractors are and for what amount they have taken on a certain job. Of course, the data is also available as open data, allowing people to analyze it themselves. Based on such examples the economic potential of Open Data is very much real. As Tim Berners Lee—widely recognized as the inventor of the worldwide web said, "If people put data onto the web [...] it will be used by other people to do wonderful things, in ways that they never could have imagined."[8]

Open Data as a Control Mechanism

Open Data does not only have economic potential. The disclosure of data by governments also creates the path for a new control mechanism on the actions of government. Politicians and civil servants can no longer easily claim—for example—that they were responsible for fewer burglaries in the neighborhood or reducing traffic. Any person can see through possible fact-free politics within a few mouse clicks. The information asymmetry between civil servant and citizen is, in large part, disappearing. Both have, at least in theory, access to the same data. The exception is information that is not made available because there is a genuine, significant reason to keep it secret. For example, data on the maintenance of armored cars is not accessible to the public for understandable reasons.

In embracing the Open Data philosophy, government is becoming much more transparent, which is in keeping with the ongoing societal change in which the importance of institutions declines.

Will that make the legislature redundant? Probably not. The Open Data concept is not so powerful that a self-governing society will be created as soon as all the data is thrown up in the air. It is, however, a huge comfort to those who lobby for the interests of consumers, privacy activists, investors or other groups. They have—whether they acquire them via the media or not— much better tools to exert control over the government on behalf of (groups of) citizens.

Limits to Transparency

Up to now, the discourse on Open Data and increasing transparency was all positive. There is also another side to transparency, however. Hospitals will have to publish their death rates; schools their exam results. They will do so because we will force them, sometimes through a media campaign. We view it as progress, because we can no longer be fooled and we have the feeling that we are now getting complete information. However, it is questionable whether such transparency will always benefit us, if only because the context plays such an important role in the interpretation of data. In the Netherlands, that became clear when attempts were made to interpret data on the performance of hospitals. For example, post-surgery complications differed ranged from 10 to 42 % for different hospitals.[9] Note that this might have nothing to do with the quality of the hospital, but with other factors related to patient population over which the hospital has no influence or control.

Making medical specialists accountable for the death rates among their patients can result in a situation in which they prefer to focus on low-risk patients and turn away complex cases.[1] It is understandable, therefore, that in the spring of 2014 the Erasmus Medical Center in Rotterdam, the Netherlands, decided on principle not to comply with new legislation enforcing hospitals to make the death rates public.

Hospitals believed that fair comparison is not possible because the differences in patient populations are not taken into consideration. Patients with heart failure, for example, are sent to university hospitals by regular hospitals. It is clear from this case that in more complex situations only specialists in data analysis can conclude anything meaningful about the quality of a hospital.

Number Fetishism

It is even clearer that the world cannot always be captured unambiguously in numbers. During debates, many politicians submerge us in numbers as if they were unshakeable truths.

The reality is that throwing numbers around—creating transparency—results in a focus on meeting targets without considering the underlying objectives. A situation is created that is focused on 'gaming the numbers'. Returning to the analogy we used in the previous chapter, a professor who gets into the 'education game' will have the urge to score as many points as possible by publishing a lot, but is soon focused on replicating his existing knowledge instead of developing new insights. Another option is to manipulate (the interpretation of) the numbers yourself. The number of patients with bedsores caused by poor care in a hospital can as easily be ascribed to injuries due to incontinence.

Another good example of the different interpretations of data that are possible is the frequently heard statement that if all Chinese and Africans had the same standard of living as us, we would run out of raw materials. According to Hans Rosling, professor of International Health at the Karolinska Institute in

Stockholm, Sweden, this is not true at all. According to him, over-population will begin to ease when we raise the living standards of the two billion poorest people on the planet to the level where they too are able to buy a mobile phone and a car. We are collectively blind to this, however, because we interpret data pretty dumbly.

In a TED talk, Rosling showed that we should be very optimistic about the growth of developing countries.[11] With amazing, visually strong and clear statistics, he showed that almost all countries in the world have made great leaps forward in both health and wealth within one or two generations. Also, some countries that seem poor now have done very well if you consider where they are coming from. He even dares to state (and he is not alone in doing so[12]) that of all the continents, Africa has performed best in recent decades. Rosling's purpose is to promote a global vision based on facts by increasing the use and understanding of freely accessible, public statistics.

Hans de Bruijn, professor of Organization and Management at the Technical University in Delft, the Netherlands, argues in his book *Measuring performances in the public sector* that the activities of organizations cannot be reduced to unambiguous statistical indicators. According to him, there are three important developments in this respect. First, everything is connected to everything else and causal connections are therefore very complex. Second, virtually all knowledge and information is controversial, almost by definition; the unambiguity of the past is simply gone. Historically, as a specialist in a certain field, you had a natural authority and your knowledge was undisputed. That certainty is gone now. Third, a huge amount of time elapses between decisions being taken and the effects of those measures occurring. Those three developments occur at a societal level as well as for individual organizations and are the main reasons why traditional models for performance reporting no longer apply. The linear forms of accountability are, by definition, a fictive reality. There are always multiple truths and whether or not a performance is good or not depends only on how you look at it. Despite this, we think that we can reduce everything to numbers and summarize them in a neat ranking.

An experiment in the city of Toronto, Canada, is a great illustration of the notion that it is sometimes best to limit transparency. The city made a change to the sequencing of traffic lights at pedestrian crossings.[13] The theory behind the change was the belief that showing the countdown to the green light would improve safety: pedestrians would be willing to wait a little longer if they knew that green was coming. The opposite was true: more accidents happened. According to the researchers, the explanation was that there was too much transparency. Car drivers were also shown how long it would be before the light would turn red. This motivated some car drivers to step on the gas, while others became more careful. That cocktail of conflicting reflexes resulted in additional accidents. The episode shows that transparency does not always contribute to a better world and sometimes has demonstrable disadvantages.

Even so, the rise of performance measurements seems unstoppable since nearly everything can be measured. The activities of the Silicon Valley startup BetterWorks give a taste of what has yet to come.

Their software lets groups of employees collaborate in setting each other's objectives. Everyone can see how everyone else is doing, by means of a smartphone app. It is called 'quantified work' and it is a big step up from conventional performance measurements. It introduces real-time performance measurement, a practice that is very common for many big Internet companies. It also has its disadvantages, which were summarized by *The Economist* as "the quantified serf" such as "managers who believe they have been set a goal that is unattainable are more likely to abuse their subordinates". Gary Latham of the University of Toronto says "it's like taking out your frustrations by kicking the dog."[14]

In our opinion, far-reaching transparency is in many cases the most obvious way to build sustainable relationships and it is an important principle in the new information society. It is, however, important to neutralize the undesired side effects, as described in this chapter, as much as possible.

Chapter 10
Systems Determine Our Behavior

But the legislature is not powerless

Slave to the Navigational System

In early 2008, a 32-year-old man followed the instruction of his car's GPS device and ended up making a right turn onto the railroad tracks, getting his rented Ford Focus wedged between the rails. The man called 911 and waited to wave the train to a stop. But the train couldn't break in time and ended up dragging the car for 100 ft until it burst into a fireball. The train's 500 passengers were stranded for two hours, another 10 trains were delayed, three trains out of Grand Central Station were canceled and 250 ft of third rail was damaged. The reason? The GPS device meant for the driver to turn right onto the northern lanes of the Saw Mill Parkway a few yards farther up, not onto the railroad tracks, but maybe some people put more faith in the inerrancy of a device rather than what their eyes tell them.[1]

We noted earlier that legislation, almost by definition, lags behind social and technological developments and therefore sometimes does more harm than good. In this chapter, we examine whether there are solutions to this. We do this, among other things, based on some examples from history, from which a number of principles emerge that are worth keeping an eye on. In the new information society, there are new ways to safeguard those principles, including developing new systems.

© Atlantis Press and the author(s) 2016
S. Klous and N. Wielaard, *We are Big Data*,
DOI 10.2991/978-94-6239-183-3_10

The example of people slavishly following in-car navigation systems shows how powerful the impact of systems on our behavior can be. That raises the question of whether we can use systems in such a way that we can dispense with the need for legislation. That is not the case and we will demonstrate why, based on a recent example where a technical subject (net neutrality) required a statutory anchor to guide the setup of the new information society in the right direction.

Kronos Effect

We start with a brief analysis of a pattern that seems to be repeated throughout history. It always starts with companies that offer more freedom and convenience based on a new technology—and it always ends with powerful parties that attempt to sabotage innovation and use their power wrongly, driven by a desire for self-preservation.

In his book *The Master Switch*, Harvard professor Tim Wu speaks of the Kronos Effect. He refers to the mythical figure Kronos who, according to the ancient Greeks, was the ultimate ruler of the universe. According to mythology, he was warned by the Oracle of Delphi that one of his children would overthrow him in the future. With that scenario in mind, he decided to eat his children every time his wife gave birth. This turned out to be his downfall. His wife fed him a stone wrapped in swaddling to save their child and he was finally defeated by his son Zeus whom he thought he had eaten.

Kronos is synonymous with large companies in the information and communications industry which historically have often become monopolists. There will come a time when they are more concerned with defending their own positions than with developing even better products and services. The outcome is the destruction of better products by competitors or 'cannibalization' of these competitors by simply buying them.

We offer two historical examples of this pattern. The first is the invention of the telegraph. Mid-19th-century America was captivated by this new medium. It promised huge freedom of communication, offering the ability for messages to be sent quickly over long distances for the very first time.

Western Union built a strong position in a short time by acquiring several smaller players. It proved to be a positive move, because the quality and reliability of the equipment of these small players often proved to be inadequate and the powerful status of Western Union led to more innovation and quality.

The growing strength of the company, however, also proved to have a darker side during the nail-biting presidential election of 1876. It was a very close call, where the Democratic candidate Samuel Tilden seemed to have narrowly defeated the Republican Rutherford B. Hayes.

However, there was a prolonged dispute over who had really won because of uncertainty about the validity of the counts. Western Union played a crucial role in the resolution of the dispute by forwarding the correspondence of the Democrats to *The New York Times*, which in turn forwarded it to the Republican camp so they could base their strategy on it. This was all in spite of Western Union's pledge that all messages sent over its telegraph system were 'strictly private and confidential'.

Tilden eventually lost the battle for the White House, partly due to the tightly orchestrated 'bugging' of the Democrats' telegram traffic. Later, it turned out that the Democrats probably were not entirely innocent either, as it has been suggested that they bribed vote counters in the crucial states of Florida, South Carolina, and Oregon. Using telegrams. Tilden, however, denied this until his death.[2] This story is reminiscent of the NSA bugging scandal and the role that large Internet companies like Google and Facebook have played in it in recent years. The Western Union story offers an interesting example of an early abuse of technological innovations. However, there are also parallels with the Kronos Effect.

The enormous power of Western Union—built on the promise of creating something great—slowly crumbled in the subsequent decades. The main reason for this was the company slowing down innovation, getting involved in shady business practices and its unsuccessful attempts to save itself by acquiring promising new businesses. It also conducted a long and, although successful, draining fight against nationalization. Western Union was eventually faced with a greater challenge that ultimately meant the end of the telegram business: the invention of the telephone.

A second example can be found in the history of the film industry. During the first half of the 20th century, master strategist Adolph Zukor kept an iron grip on the film world through his ownership of stakes in not only filmmakers and distributors, but also cinemas themselves. What is now known as vertical integration made it possible for him to dictate the what, when and where of film screening. As he had so many stars under contract, he could also force cinemas to acquire substandard productions through a system of fixed packages, called block booking. When Zukor's company, Famous Players-Lasky, wanted to merge with First National—thereby becoming even more powerful—Zukor overplayed his hand. A group of major movie stars, including Mary Pickford and Charlie Chaplin, decided to join forces and organize their own independent production company. They said they were doing so to safeguard their artistic freedom, although it was obviously also good for them financially. Zukor faced lengthy lawsuits over the use of block booking, which was finally banned altogether in 1948,[3] bringing the vertical integration of the film industry to an end.

The Power of Google

Back to today. What does the Kronos Effect mean for today's Internet industry? It is an accepted fact that players such as Google, Facebook, Amazon, Apple, and other Internet giants have a powerful position. Google in particular is often cited as an example of a company with too much influence over everything we do. This reflects not only the size of the company, but also its wide portfolio of products and services. Well-known services like Gmail, Google Maps and YouTube, but also relative newcomers to the broadcasting family, such as Nest (smart thermostats) and Deep Mind (artificial intelligence) fall under the Google umbrella.

The similarities with Western Union in a number of areas are striking. Just as Western Union promised the world significant change for the good, Google's ethos reflects a certain idealism, namely the desire to disclose all the information in the world. From its inception, the company operated based on its corporate slogan 'Don't be evil', and few would dispute that Google has indeed brought the world many good things.

Without a doubt, Google will go down in the history of the Internet as a company that achieved many great things. The question is whether, in the course of that history, Google will join the list of companies that fell prey to the Kronos Effect. Some critics believe that Google has not lived up to its slogan for years and wants more power first and foremost. It is also true that the company—just like Western Union—is making numerous acquisitions. The key question is whether Google is doing this to serve us better or for reasons of self-preservation. This question has also entered the political agenda.

In an interview with the *Frankfurter Allgemeine* newspaper, German Minister of the Economy and Energy, Sigmar Gabriel, hinted at the company being forced to break up if new competition rules for Internet platforms did not help to prevent Google from "systematically crowding out competitors."[4] In the autumn of 2014, a motion was adopted in the European Parliament mentioning the breakup of Google as one of the options to ensure healthy competition in the Internet industry.[5] Of course, the European Parliament does not have the authority to enforce any such breakup, but it does show that the pressure is increasing. The reaction from the US was that the motion was foremost a political move and not substantive.

The question of market dominance applies not only to Google, but also to other Internet giants. A striking example is the acquisition of WhatsApp by Facebook in 2013 for the unprecedented sum of $19 billion. As far as we know, it seems clear that it will be nigh on impossible for Facebook to recoup this investment in the short term because WhatsApp does not even have a proper revenue model. It would appear, therefore, that Facebook bought the service to protect its own strategy—connecting as many people as possible—against new competitors. This is also reminiscent of the ultimate ruler Kronos eating his newborn offspring to defend his throne.

Preventing the Abuse of Power

Can we prevent such abuses of power by companies? The issue is, of course, far from new. Governments worldwide use tools such as competition laws and supervisory authorities to combat these abuses, with the underlying goal of ensuring that consumers have optimal freedom of price, quality and choice. These tools have their limits, however, with one of the main additional complications being that there are many free services on the Internet.

One of the consequences is that many of the big players in the Internet industry would fail the 'hypothetical monopolist' test, which considers whether a hypothetical monopolist could raise prices by a small but significant non-transitory amount, such as 5 or 10 %, because 10 % of zero is still zero. This requires a different perspective on antitrust rules. Two American experts, Allen Grunes and Maurice Stucke, mention in an article[6] that a data barrier to entry exists and that it hampers competition from startup search engines and social networks, among others. They refer to the case of US versus Bazaarvoice, which involved the completed merger between the two largest providers of online ratings and reviews. In discussing the entry barrier documents, the court highlighted a document prepared by Bazaarvoice for the investors' roadshow before its IPO. Among other things, this document talked about the company's ability to "leverage the data from its customer base" as "a key barrier [to] entry". When the matter came to trial, Bazaarvoice tried to walk away from these characterizations, saying it was really talking about the company's competitive advantages, and real economic barriers were minimal. The court disagreed: "Much of what Bazaarvoice refers to now as its 'competitive strengths', it used to call, accurately, significant barriers to entry."

We mentioned earlier that the strategy of many Internet companies is currently mainly focused on obtaining more sensors from cell phones to refrigerators that are connected to the Internet and allowing them to map our behavior. The data they collect is central to their service offering and also enables them to develop new products and services. Such extensive data collection creates a risk of the abuse of power. Companies like Google have such a big lead in terms of the volume and variety of the data they are collecting that they are becoming untouchable. The lead is so great that for a newcomer to the market to be able to compete with Google seems hopeless.

Legislation to prevent the abuse of power by companies like Google remains limited. Maybe we need laws that impose limits on the volume or the depth of data collected, instead of the traditional approach of limiting market share. Whatever the answer, this issue confirms that legislation is almost always playing catch-up.

This does not exactly mean that we cannot do anything about the abuse of power. We should bear in mind that history—as in the case of Western Union and other examples—shows that when a powerful party overplays its hand, it is usually the beginning of the end. There is no reason to assume that this has changed. If one of the powerful Internet companies crosses the line, the consequence will, without a doubt, be that its own position is undermined. It may even lead to a forced breakup to contain its power. The impetus for corrective action would be especially strong nowadays as transparency is so much greater and citizens are empowered. We described earlier how the power of the collective can force companies to change. If you don't listen to the collective, you have a problem. The question that remains is whether legislation is required.

Cyberspace as an Independent Republic?

Discussions on how to legislate for and regulate the Internet have intensified in the last decade as the Internet has matured. Many of the very first Internet users dreamed of an Internet— or more specifically, activities on the Internet—which would be unregulated. They envisaged the whole of cyberspace as their independent republic where governments had no influence. The following is an extract from the beautiful 1996 *Declaration of the Independence of Cyberspace* by the cyberlibertarian movement headed by John Perry Barlow: "Governments of the Industrial World, you weary giants of flesh and steel, I come from Cyberspace, the new home of Mind. On behalf of the future, I ask you of the past to leave us alone. You are not welcome among us. You have no sovereignty where we gather."[7]

Some 20 years later, we know better. Even Barlow himself knows better; in an interview in 2004, he said "We all get older and smarter."[8] Cyberspace is not a free republic (anymore). It has become part of the mainstream world where laws and regulations play a major role. Gradually, however, something else has also become clear. In cyberspace, software and hardware increasingly determine our behavior, probably to a much greater degree than laws and regulations.

We began this chapter with a couple of examples of how navigation systems affect what we do and how sometimes that does not end well. We did so for good reason. Such control of behavior by systems arises all the time and is fed by data. Online, it is not the law, but rather Google's algorithm that determines which items are relevant to you. It is not the law, but the software and hardware of the smart card system that determines how you use public transport. 'The system' is becoming increasingly dominant in determining our behavior.

That in itself is no bad thing because we can change the code, meaning our systems, to our liking if we want. However, we need to make an important point. If a system is the equivalent of the law, the system developer must be the legislator. Moreover, the influence of a system on behavior can be even more prescriptive than that of legislation because the law only specifies how we should behave. In some cases, a system goes beyond this and sets hard limits on how we can behave.

Again, we can use the public transport smart card system as an example. If you want to use public transport, you will have to do so according to the conditions the developers have embedded in the system.

If systems are to exert such a strong influence on our lives, it is essential that their architecture should be aligned with the principles underlying existing legislation and with what we want as a society. This is often where things go wrong. The principles that we adopt in society—for example, in the areas of privacy, transparency, and 'the right to be forgotten'—are not the top priority for system developers even though they should be. If they were, the technology could be used to provide a much higher level of protection for our fundamental rights, including privacy.

Net Neutrality

So far, the conceptual discussion makes it clear that despite the fact that systems are increasingly determining our behavior—or perhaps because of it—legislation remains necessary. How the balance between social needs, legislation and systems looks in practice became apparent recently in the debate around net neutrality. Net neutrality is a principle that is closely related to the anti-trust legislation mentioned earlier. It means that Internet service providers enable access to all content without distinction and do not prioritize or delay any Internet traffic.

Net neutrality is really a theme that is not subject to geographical boundaries and therefore it merits international implementation. The European Parliament bill of 18 March 2014 made it possible for Internet service providers to prioritize 'specialized services'. This means that a provider may not block access to a service, but may impose extra charges for an 'improved quality of service offering'. In theory, this will open the door to different commercial offers—or an Internet with different speeds—although opinions differ on the precise interpretation of the bill.[9]

To many people, net neutrality sounds boring—'nothing to do with me'. However, this is exactly the sort of important area that needs a solid legal basis, with parallels, in some respects, to the right to free speech. We really need to recognize that something major is at stake here. This should be widely discussed in schools, if only because companies already recognize its importance. It is telling that EU Commissioner for Digital Agenda at the time the legislation was passed, said that she was "buried under lobbyists" regarding the debate on net neutrality.

It's not easy to convince everyone of the importance of net neutrality. It is reassuring that there are other ways to attract the attention of a large audience to subjects that are "as boring as a shipping container", as it was described by the British comedian

John Oliver during a major rant about it on his weekly show on HBO.[10] His entertaining plea even resulted in a minor revolt. Following Oliver's call to action, tens of thousands of people visited the Federal Communications Commission's website, with the huge traffic temporarily shutting down its servers.

Conclusion

Net neutrality is a fundamental principle that may benefit the new information society as a whole and that, at a minimum, definitely contributes to healthy competition between companies. Such a principle should be a fundamental part of the structure of the new information society; legislation and public awareness play a crucial role in this respect. We cannot simply hand the responsibility for embedding the principle in the structure over to the system developers and data scientists. In the next and final chapter, we will therefore outline an ecosystem that places the appropriate responsibilities on the appropriate parties.

Chapter 11
A New Ecosystem

The new information society requires new assurances

The History of Accountancy

In the 19th century, Lodewijk Pincoffs was an eminent business-man in the Dutch city of Rotterdam. One of his companies, how-ever, was in turmoil. Balance sheets were massaged and records falsified to hide the fact that business on the west coast of Africa was bad. Instead of a profit of 2 million guilders, as the balance sheet stated, there was a loss of 8 million. At that time, it was an accounting fraud of enormous proportions. The two bookkeepers responsible for the balance sheet were not independent, which was very common in that time, according to the book History of accountancy in the Netherlands *by Johan de Vries (Van Gorcum 1985). It was completely in the spirit of the time to not audit the books and to stay silent about dirty dealings. The affair is widely recognized as the birth of modern accountancy. The necessity of reliable financial information became immediately relevant due to the affair.*

Information only becomes valuable when we know how reli-able it is. That applied to the accounts of Lodewijk Pincoffs' com-pany Afrikaansche Handelsvereeniging, but still applies to the new information society. Moreover, assurances about reliability are more important than ever. This applies not only to the relia-bility of the information itself, but also to compliance with agree-ments on how information is handled.

© Atlantis Press and the author(s) 2016

S. Klous and N. Wielaard, *We are Big Data*,
DOI 10.2991/978-94-6239-183-3_11

We want to be sure that a medical specialist makes the best possible diagnosis based on the proper data. We want a self-driving car to be fed with reliable data and the algorithms used to guide us safely through traffic. We want a guarantee that we take the right financial decisions based on data analysis. We expect a new app on our smartphone to be transparent about how it handles our data and we want to make sure that our personal data is treated confidentially and not used for purposes we disagree with.

A lot is at stake because Big Data is rapidly conquering society. That is a cause for some concern because the list of examples of where it went badly wrong in recent years is pretty long. Organizations are not transparent about what they do with data. They do not take the privacy of people seriously and, in some cases, they use ill-considered applications for data analyses.

This is largely because there are no standardized solid structures in place for the responsible use of data. In fact, organizations are continuously reinventing the wheel in a technological, organizational and ethical sense. Only one conclusion is possible: we have to handle data (and Big Data) differently. We need a new foundation. The question is how serious a data incident has to be—as bad as the Pincoffs affair?—before we start laying a solid foundation under that new information society. We hope that a proper debate will create sufficient awareness to be able to avert such drama.

A New Ecosystem

What we need is an ecosystem in which parties such as governments, companies, programmers, data scientists, stakeholders, lawyers and users can best work together on good Big Data applications. In this ecosystem, the parties that own data—from telecoms companies, banks, and governments to Internet companies—can obtain insights responsibly and make these insights available in the right manner. The cliché says that sharing knowledge equals multiplying knowledge and that is undoubtedly true, especially in this context because data analyses are particularly valuable when they combine insights from various sources. For instance, think about the leap forward that can be made in improving medical diagnoses and treatments. That leap will only be possible if we are willing to provide insights into our (anonymized) data for research that will benefit our fellow humans.

Therefore, there is a need for an ecosystem in which like-minded parties can meet and where they can develop applications very easily and efficiently, without constantly reinventing the wheel. It should be an ecosystem that also motivates participants to develop the best possible applications, and that offers assurances that will address the legitimate concerns of people where the use of their data is concerned.

Such an ecosystem should become a hub in which the insights from many sources come together, a hub in which anyone can participate. However, it should also be a hub where there is no misunderstanding about the conditions that apply if someone wants to generate or use those insights. This does not only apply to fees (e.g. in the form of a subscriptions) but also to the manner in which insights are created and for what purposes they can and cannot be used. There are various options to ensure that data is handled responsibly. These options will vary depending on the circumstances.

An example to make that clear: a mobile phone provider can collect information in real time, based on GSM signals, on how many tourists are at the seaside on a sunny day. Those insights can then be used—whether or not for a fee—in other data analyses, such as to predict traffic flows later that day. If data is used in this way, it is unlikely to face much resistance.

There are, however, situations in which resistance may be expected, and more assurances are required. These assurances could for instance make it necessary for two owners of sets of data to provide their insights to a Trusted Third Party (TTP) acting as a notary. The participating parties would not obtain any insight into each other's data, but it would still be possible to provide services based on the combined data. One example of this is the previously mentioned device that monitors driving behavior for an insurance company. In order to safeguard the privacy of the insured, collecting data on driving behavior and the related analyses can be performed by a party other than the insurance company. A TTP can then combine the outcomes of these analyses with other information about the insured (such as age, claims history, etc.). Only when it is necessary will information be provided to the insurer (and the insured), such as if dangerous driving behavior has consequences on the driver's premium.

There are also situations where stronger assurances are required, in particular in health care. It can be arranged for health-related information to only be available at approved locations (at health care institutions or in the files owned and managed by the patient), and for any analyses performed on that data to be done at the same location. This method can be traced directly back to the way we perform analysis in recent years at CERN. The analysis is distributed across several data centers.[1] Its essence is that we do not bring data to the place where the analyses occur, but turn it around: we send the analyses to the data. For this purpose, the analyses can be split into subanalyses that are performed at the location of the owners of the data itself. Combining insights from each party is often only required for a small part of the analysis and can be done on derived or aggregated data. That part of the analysis can even, if required, be performed by a trusted third party. The main benefit of this method is that the owner of the data retains control of the analyses that are executed and does not have to allow any data to leave its location. This is especially relevant for medical data, telephone data, or financial data. The PopMedNet software is a set of tools that makes exactly this possible: "In health care and other fields there is often a need for institutions to collaborate by sharing specific information, but a strong reluctance to share large volumes of sensitive or otherwise protected data. [...] PopMedNet overcomes these distributed querying barriers through use of flexible governance mechanisms and a simple architecture that keeps the power in the hands of the data holders."[2]

No Doubt About Privacy

Using these methods, we can also meet one of the key preconditions for a responsible application of Big Data: privacy. For many years, there has been a heated debate on privacy, resulting in various new laws and legislation. As mentioned in the previous chapter, that is only really of benefit if we are able to translate the privacy principles into systems that can enforce that privacy. Compare it with a subway. The law requires us to buy a ticket, but the system of access gates determines our actual behavior. The legally embedded privacy principles can be enforced in the same way in the new ecosystem using a system of assurances. As in the case above, one of the options is to use a TTP that can anonymize data, which can therefore no longer be traced back to individuals. This makes data analyses possible, without making personal data available. The system enforces privacy.

History shows that being diligent when anonymizing data is far from an unnecessary luxury. In the US in the 1990s, privacy was not given sufficient consideration when large quantities of medical data were released for medical research. Anonymizing data was accomplished by simply filtering personal information from the data. Later, however, it was demonstrated that by combining various data sources, re-identification (retracing data back to individuals) was very simple. Latanya Sweeney, a graduate

student who has since gone on to become a Harvard professor of Government and Technology, proved that in a high-pro-
file court case by finding (with little difficulty) the medical data
of the then Governor of Massachusetts, William Weld, who had
just previously become ill and collapsed during a public appear-
ance.[3] Anonymizing data requires more than filtering personal
information.

For many of us, it is perhaps difficult to understand exactly
how our privacy can be protected. Let's look again at the example of the bank that can block your credit card as soon as it is
used somewhere where your mobile phone is not. This can, of
course, increase payment security, but it is to be hoped that the
bank and the telecoms provider will not be willing to simply hand
over without question the personal data that is necessary for this
application.

This problem could be solved if we asked both the bank and
the telecom provider to anonymize the combination of ZIP code
and date of birth using the same encryption key, with the help
of a TTP if so required. The process would work as follows: the
bank keeps a table in which this anonymized data, the pseudonym, is linked to the GPS coordinates of the location where the
credit card is used, the telecoms provider does the same and
creates a link in this table between the anonymized data and the
location of the mobile phone. This may sound complicated, but
technically it is child's play and as a user you do not notice it.
When you use your card, the bank sends a request to the telecom provider within a fraction of a second. The provider sends
the encrypted location of the mobile phone that belongs to your
pseudonym to the bank. The bank compares the encrypted location of the phone with the encrypted location of the credit card
in order to determine whether the phone that belongs to your
pseudonym is at the same place as where you want to pay. If the
locations are the same, the bank approves your payment. If they
are not, the bank still does not know where your phone is, but it

does know that it is not near the location where the transaction is happening. The bank can then block your credit card to prevent fraud.

In this example, the bank does not use more than the minimum amount of information required to meet the objective, safe payment traffic, and it does this by using anonymized data. The information that parties share as a result of their analyses cannot be traced back to persons or locations and is, in fact, worthless for any application other than what it is meant for, which is preventing credit card fraud. The system simply enforces the responsible use of personal data.

A Review Model for Data Analyses

The ecosystem does not only have to ensure privacy, but also offer guarantees in other areas. We concluded earlier that data analysis is much more than number crunching and that the quality of complex analyses in particular is not determined by the computer. As we made clear before, the data scientist has to have the analytical capabilities of Sherlock Holmes. The difficulty lies especially in being able to understand or interpret the results, without having any preconceptions. In other words, we are looking for a mechanism that encourages developers to strive for the highest possible quality of data analysis.

In the world of data science, interesting trends are occurring in this respect. It is a relatively young field and therefore the professionalism of the working method of a data scientist is not a given. Big steps have been taken to improve this in recent years. Open standards such as the Open Data Platform ensure that the recyclability of analyses is increasing, which in turn increases their quality. Furthermore, platforms for data science are becoming more professional due to the integration of old friends from the world of software development such as tools for sharing knowledge or managing source code. All this encourages the data scientist to work in a more structured and efficient way, but encouragement is not enough.

Since data analysis is increasingly penetrating society, it is necessary that the quality of the applications themselves also be guaranteed. The keystone of quality assurance is a review model for data science, where not only users should be able to give ratings and reviews in order to share experiences, such as in an app store, but data scientists should also review each other's work. Perhaps specialized parties will start to certify those analyses. This may result in a mechanism that promotes the quality of the underlying data analysis through a dynamic comparable to an app store.

This may all sound farfetched, but that is not the case. More to the point, some of it already exists. The website algorithmia.com, where a market place for algorithms is being developed, is just one example. "Users can create, share, and build on other algorithms and then instantly make them available as a web service."

We foresee an ecosystem in which elements of the familiar App Store will appear, useful or fun applications will rise to the surface 'by themselves' and others disappear into the background because of a review model where users provide assessments and reviews in order to share experiences. The same system is possible for data analyses. An application for predictive maintenance for cars could perhaps be recycled in other sectors because the algorithms are similar. An application for monitoring and checking invoices that works well in one energy company may also be suitable for another energy company, making it possible to share the development costs. When both users and specialists review each other's work like this, the best applications will rise to the surface by themselves.

The comparison with the App Store only applies to a certain extent, however. In this case, it is not only about the usefulness or the pleasure someone gets from an application, but also and even more about the substantive correctness and the responsible application of the analysis.

Quality Control

The ecosystem cannot function without properly enforced quality control. The stimulus given by the application users' review system is not enough, because it has no visibility and therefore can give no feedback on how data is handled 'behind the scenes'. In addition to the user reviews, there should also be a role for assessment by a specialist who is not focused on convenience, but on issues such reliability and quality. While this may sound a little abstract, it is already being used here and there in consumer product comparisons. In these cases, web shops combine insights from user reviews and specialists. Actually, it is remarkable that such combinations are hardly ever used to review for example hotel rooms or restaurants. Those review websites currently focus rather one-sidedly on user experiences.

This is very undesirable in the new ecosystem we envisage, simply because there is too much at stake to leave the quality assessment only to the users. In order to achieve the best of both worlds, it is necessary to add a professional assessment by subject matter experts that operate independently, have in-depth knowledge and a reliable reputation.

Historically, that role was filled by accountants. Perhaps they can be rehabilitated and pick up where they left off? That thought is not entirely new. Robert Elliot, the former chairman of the American Institute of Chartered Professional Accountants AICPA, at the time of accepting his position, around the turn of the century, said the following about using information, "Traditional accounting does this by enabling better investment and managerial decisions to be made. But I am talking about a much wider range of information leveraging. The basic ingredients would be broad business knowledge, analytical power, and expertise in what I'll call 'knowledge science'."[4] At that time, the term Big Data was not yet being used, but even then, Elliot foresaw a scenario in which the accountant would act as quality controller in the new information society.

This has not happened in the last 15 years; accountants have not yet made this innovative step forwards. This is possibly because of the credibility problems that have followed the sector for years now because of a series of (accounting) fraud issues.

Innovation can and will also occur outside of the traditional domain of the accountant. We refer to the growing need for insight into the impact of a company on the environment and society, preferably at the level of a product. This is also the background of the development towards integrated reporting, in which the financial value is connected to societal impact. At product level, the question is how much profit a company makes on the sales of (for example) a pair of jeans, once the damage to society and the environment is taken into account. Although this is not easily determined, because complex connections are sometimes involved, Big Data provides us with increasingly better possibilities to make these analyses.

The examples mentioned comply with the quality control requirements of the ecosystem described. The related innovations are not gradual, but groundbreaking concepts that contribute to the other basic principles we mentioned that are of social interest in the ecosystem, such as privacy, true value and the sustainable financial situation of companies.

The New Ecosystem Is Currently Developing

The new ecosystem is perhaps closer than we think. The Internet has three characteristics that align seamlessly with the demands we will make of it. First, there is redundancy: there is no single individual party or component that can 'switch off' the internet. Second, there is reduction of complexity. The network itself contains very little intelligence and communication runs through relatively simple components that follow simple rules. The intelligence is found in the 'end points'. Third, there is modularity. Every component or party is responsible for a relatively small part of the larger whole. None of the parties should be irreplaceable.

There are practical signs that the ecosystem can be created. A striking example is a do-it-yourself initiative from Jesper Andersen and Toby Segaran, two mathematicians. Their initiative is related to credit ratings for companies.[5] They started an open source project in reaction to a much-criticized structural flaw in the existing credit ratings system. This flaw is that the agencies that currently determine the ratings are being paid by the companies that receive the ratings. The central question that came up, especially after the financial crisis, was whether those agencies are independent enough. The project by Andersen and Segaran, freerisk.org, does not have that flaw. Thanks to this project, anyone with a calculation model can analyze the enormous amount of SEC filings (the information that listed companies have to deposit with the American stock exchange supervisor) about a company. If this works, and anyone can DIY the credit rating of an obligation or company, it could potentially negate the need for credit rating agencies.

Not only can everyone using their own model assess solvency, but that model can also—completely transparently, contrary to the traditional manner in which ratings are made—be offered to co-users. The idea is that in time the best model will rise to the top by itself.

Admittedly, freerisk.org has not turned the financial world upside down—at least not yet. The traditional method of three influential credit rating agencies making more or less invisible calculations still exists. The line of thinking is at least interesting, however. When we disclose data in a simple manner to everyone, the best applications will come out on top by themselves. Perhaps freerisk.org is ahead of its time.

Now that the information society is maturing—and therefore also the awareness that the power of data analysis is growing—it is necessary to develop a model that provides better assurances about the quality of the data analyses.

Platform Thinking

Previously we discussed two issues that underline the potential success of such an ecosystem. The architecture of the Internet is very suited to it and DIY communities are being created that have many of the characteristics of such an ecosystem.

There is a third argument. The Big Data ecosystem directly connects to a trend that in recent decades was the foundation of many online successes: platform thinking. The most obvious example is the enormous impact that Apple has with the App Store. And of course, Facebook is an example. That platform is more than a connection between over a billion people; it also allows users to install applications created by others.

A more recent event was a much-discussed blog post by the founder of Tesla, Elon Musk, with the title *All Our Patents Are Belong To You*.[6] In this blog post, Musk announced that Tesla is heading in another direction. It will no longer strongly protect patents on electric car technology, but will allow everyone— under certain conditions—to use them. The reasoning is that in the end, this will benefit the entire world. That sounds rather philanthropic, but it also has a commercial side. History has shown that such a strategy is often applied with new products and services and is even a precondition for (long-term) success in completely new markets. This was also the economic mechanism during the industrial revolution, as evidenced by the manner in which the mechanical loom conquered the textiles market by freely sharing ideas and techniques.[7] The essence of this principle is that only when a market is developed and mature does it make sense to safeguard your intellectual property.

Take, for example Apple, which was not at all concerned with protecting patterns during the early years, but later ended up involved in full-blown wars about intellectual property, including with Samsung. Musk understands better than anyone—especially because of his earlier experience when establishing PayPal—that he first needs to create a large and well-functioning platform for the electric car before a serious business can be established. His first concern is that platform, to which everyone can contribute. Only then will the desired economies of scale develop.

Platform thinking got a lot of attention in the last decade, both in practice and from a scientific point of view. Many of the themes in this book are examples of platform thinking. Developing a platform to achieve social value is one such example, as is the simplicity with which components clicked into each other, a review system that 'democratically' encourages quality, an open design to which everyone is invited to contribute (and from which everyone receives advantages), re-using insights, and more.

Our vision of how the ecosystem should look is more than the dream of two Big Data enthusiasts. It is more the sum of all the movements that are already there. The ecosystem is already being built. If we want to steer things in the right direction, we have to consider the characteristics of a good platform strategy, to ensure that the ecosystem will be attractive.

In this network society there is only one substantial question for companies: how do I contribute to others creating value? When answering that question, platform thinking is the foundation and three conditions are essential. First, it has to be easy to make connections. Second is the so-called gravity. To what extent does the platform attract all participants, from manufacturers to consumers? Third is the 'flow'. How does the platform stimulate exchanges of information and development of value together?[8]

If we are capable of building a responsible ecosystem that meets those conditions, then we will have real progress. We can then propose answers to the problems Big Data is currently posing and make sure that all the participants within the ecosystem play a part in this respect. Is that a brave new world? No. But it is the start of a new era in which we as a society and businesses deal with data fundamentally differently from that which we are accustomed.

Finally

In this book, we discuss Big Data positively. We are convinced of its potential and therefore we want to obtain insights to control its disadvantages as much as possible. However, we cannot avoid Gartner's Hype Cycle, which is known to everyone working in the technology sector. For many years now, this agency has been using a model to show that new concepts often rapidly face overheated expectations that result in disillusionment, and only translate into actual added value many years later. To us, Big Data is, of course, much more than technology. But in 2013, it was at the top of the Hype Cycle. This can only result in disillusionment in the ensuing years. At the time of writing this book, Big Data could no longer be found on the Hype Cycle. Nevertheless, we believe that the long-term results of its derivatives could still be very satisfactory.

There is a clear analogy with the rise of the Internet around the turn of the century. The expectations were as high as the sky. Making profit was not relevant; it was all about top line growth. Gurus talked about the coming of a New Economy (in capitals) in which growth was never-ending. The construction collapsed with a bang, however, the stock exchange imploded and creditors were left with receivables by large numbers of bankrupt companies. The disappointment was huge, as was the surprise at how we could have been so naive. "We must have smoked something bad", a senior manager from a telco whispered to one of us off the record at the time.

And yet, 10 years down the line, we know that our expectations were not all that crazy. Actually, we have only recently learned how to use the Internet properly and how it can bring social progress. In fact, the Internet probably has had a much larger impact than was predicted even at the peak of the dot com hype.

Something similar will probably apply to some Big Data related applications. Many advisors are smitten with it as if it will solve all the problems society and business face. Managers want to use it without knowing what they are talking about. In fact, all the signals are on red and the hype is about to implode, indeed it is probably already happening. Take, for example, the discussions on the Big Data backlash such as in *The New York Times*.[9] This is necessary and maybe even helpful, just like the bursting of the dot com bubble was useful in the end. The implosion will create the space to set up the structures and ecosystems in the coming years so that we can fully exploit the advantages of Big Data. A quiet revolution is probably easier to achieve than one with lots of noise. But the revolution is coming, have no doubt about it.

Epilogue

The Art of Looking Sideways

"It's hard to make predictions, especially about the future" is a quote attributed to Niels Bohr, Nobel Prize winner for Physics. And, of course, that also applies to the vision we have described in this book. That vision is based on a deep-rooted conviction that we have to do things differently; that society deserves better Big Data applications than are all too often in use at the moment.

Not only does it have to be different, it can be different. We have demonstrated that as clearly as possible, and given you the reader a glance into the (possible) future. But we are only human; the new information society could turn out to be completely different from what we have discussed in the previous chapters.

Let's go back to our earlier conclusion that data is the new gold. With proper applications, we can make our lives a whole lot better and solve many social issues. If this is really so, we have to handle this data very carefully. We already said in our closing statement that we need a new ecosystem with clear responsibilities and assurances, in order for society to trust the insights that are obtained by data analyses, how these insights are applied, and the manner in which individual privacy is safeguarded.

© Atlantis Press and the author(s) 2016
S. Klous and N. Wielaard, *We are Big Data*,
DOI 10.2991/978-94-6239-183-3

This is vital because our trust seems to have been destroyed. Whom can we still trust in a world in which we are disappointed every day by the actions of companies and governments? We are shocked by the revelations of Edward Snowden on the bugging by security agencies, there is national outrage when banks want to start a pilot project to do new things with our customer data and some people even verbally abuse bankers. We are disappointed in a legislature that is not capable of organizing the proper use of data, and in the accountants that should be supervising that use. And in some cases, we don't even trust voting machines to do what they should.

In this book, we proposed ways of dealing with this reality successfully. We suggested changes to the education system to better prepare people for life in the new information society. We suggested new systems to ensure that data analysis is performed responsibly, and to avoid accidents caused by computers making autonomous decisions. We suggested an ethical debate on how we can ensure that data analysis brings good things to society and canalizes bad things. We discussed how we could create new equilibria by introducing smart mechanisms based on data. We made suggestions to prevent individual parties obtaining too much power in the new ecosystem. All these suggestions have a central purpose: to make sure that we trust the outcomes and create reliable applications even in an unreliable environment.

We assumed that this is the only way the new information society will result in a great new world.

That is true, right?

Or is it not necessary at all to create these assurances? Would the information society also function perfectly well without trust?

No Actions, But Words

A significant development in the history of the financial sector is the establishment of cooperatives. In the nineteenth century, Friedrich Raiffeisen (after whom Rabobank was named, the Dutch multinational bank that is to this day a global leader in agri-financing and sustainability-focused banking) was the founder of the first agricultural credit bank, a bank that was owned by (wealthy) farmers and whose focus was on doing good for others in society. The insurance sector has a comparable history. The first insurance company was a group of farmers that agreed that they would support each other if one of the participants were struck by a calamity, such as a barn burning to the ground. Trust in the good intentions of the participants (including the owners) was the basis of both ideas.

Compare that to our current level of trust in the banking sector. After the start of the credit crisis, it fell to zero. New initiatives to change banks are almost always received with huge amounts of cynicism and suspicion. The suggestions that these plans are meant to do right for customers or for society do not stand a chance. The managers responsible have been depicted for many years as fat cats that no longer consider the interest of the customer. Although banks themselves say that they work hard on programs with customer's interests at their core, and want to regain their trust, this rarely results in successes. The trust in the financial sector seems to be gone for the long term.

However, the banking and insurance system is still functioning despite the distrust and despite the fact that the financial crisis uncovered the fundamental flaws in the system. Hordes of people are not rushing to take their money out of financial institutions. The majority of loans are still made by banks. Insurance companies are having a difficult time but not failing. We are collectively angry at the entire financial sector, but we only convert that anger into actions to a limited extent. We do not know exactly why this is so, but the inconvenience of switching to possible alternatives will definitely be a factor. In any case, our behavior shows that we still give the banks a central role in dealing with our money, despite the distrust.

Will large Internet companies follow the same path? Almost all the technology giants of today started out really small, with a good idea about how to improve the world. Almost 10 years later, these companies have huge numbers of daily users that do not, or hardly, see the good intentions of those first days. Google and Facebook are regularly subject to severe public criticism because of the way they handle data. Google is getting as close to the 'creepy line' as it can. That strategy was explained very well by a columnist on the technology website zdNet:[1]

Google wants to control our photos, our social media, our e-mail, our text documents, our task lists, even our common documents. It wants to harvest that information and feed it to the Borg that is by far the world's largest information filtering service. As long as we keep supporting Google's model, we are feeding the centralization of human knowledge, e-mail by e-mail. This is the brave new world of the Internet, where privacy is a historical footnote and where we are fooled into giving it up or just bought off.

In 2006, Facebook already crossed a virtual line in the sand with a new advertising program called Beacon. This program makes it possible to record the online behavior of users outside of Facebook and to use that information for ads directed at the friends of these users. After a hail of criticism, Facebook retracted the program. In the following years, many more incidents occurred. Time and time again, there is a lot of anger about how Facebook deals with data. Its stock market launch in 2012 fed that resentment further by making it painfully obvious that billions were being made using data that we as users still consider our property. The level of trust in the way the company handles our data is therefore very low, and that does not only apply to Facebook, but also to many other companies whose commercial operations are based on the use of (our) data.

Despite that lack of trust, the use of these services is still growing and users are still storing data there. We do not know exactly why, but the inconvenience of possible alternatives will definitely also be a factor in this.

Could it be that it will continue like this even in the longer term? That we—as we have seen in the developments of banks—will consider Google the easiest way to manage our data, even though its original good intentions are no longer apparent and incidents that erode our trust occur regularly? That we will use Facebook and Google, or their successors, to store data on our diets, our health, our finances, our friends, our consumption patterns and our hobbies? That trust does not matter?

Why not? The future will probably be more absurd than we can imagine. And maybe it's not so absurd after all. We let the businesses we trust the least handle our money. It is likely that we will also store our data with companies that we trust the least. If that is the case, you could have saved yourself the trouble of reading this book. And apologies are in order. Sorry.

Word of Thanks

Writing a book together is great fun. In a few sessions at small plastic furniture without Wi-Fi—Sander was in the middle of renovating his new home—we built a storyline with some key phrases. Soon, we were bouncing up and down with enthusiasm. Then the real work started; for months, we bombarded each other with scientific material, wandering thoughts and news reports. We threw away half of it and of the other half formed—in our minds—a logical story. The storyline then turned into a book.

During this process, we were supported by many people that were willing to share their insights, to edge us toward hitherto unknown paths or give their blunt opinions. Therefore a special word of thanks to Mirjam Bult, Arjan de Draaijer, Leon de Jong, Cees de Laat, Viktor Mayer-Schönberger, Marcel Pheijffer, Egge van der Poel, Eric van Rooijen, Erik Schut, Pim van Tol, Brenno de Winter, Sandra Wouters and Gerrit Jan Zwenne.

© Atlantis Press and the author(s) 2016
S. Klous and N. Wielaard, *We are Big Data*,
DOI 10.2991/978-94-6239-183-3

Notes

Introduction: Anything Is Possible

1. https://www.youtube.com/watch?v=2vXyx_qG6mQ
2. http://en.wikipedia.org/wiki/Creative_destruction
3. http://www-01.ibm.com/software/au/data/bigdata/
4. https://en.wikipedia.org/wiki/Quantified_Self
5. http://www.peterhinssen.com/books/the_new_normal

1. Big, Bigger, Biggest Data

1. http://www.economist.com/news/briefing/21582042-it-getting-easier-foresee-wrongdoing-and-spot-likely-wrongdoers-dont-even-think-about-it
2. http://www.sfweekly.com/2013-10-30/news/predpol-sfpd-predictive-policing-compstat-lapd/full/
3. http://abcnews.go.com/Technology/software-predicts-criminal-behavior/story?id=11448231
4. http://www.pcworld.com/article/2032913/the-end-of-oores-law-is-on-the-horizon-says-amd.html
5. As also observed by the Dutch professor in systems and network engineering Cees de Laat: https://twitter.com/cdelaat/status/327479877689307136
6. https://gigaom.com/2013/09/18/attacking-cerns-big-data-problem/
7. http://www.gartner.com/newsroom/id/2636073

8. http://www.nytimes.com/2012/02/12/sunday-review/
    big-datas-impact-in-the-world.html
9. http://link.springer.com/article/10.1007/s100529801030
10. http://ec.europa.eu/transparency/regexpert/index.cfm?do=gro
    upDetail.groupDetailDoc&id=2245
11. http://www.worldclassmaintenance.com/nl/component/tags/
    tag/3-windturbines
12. https://www.teslamotors.com/nl_NL/presskit/autopilot
13. http://www.forbes.com/sites/bruceupbin/2013/10/02/monsa
    nto-buys-climate-corp-for-930-million
14. http://gigaom.com/2013/11/02/venture-capital-in-an-age-
    of-algorithms
15. http://www.nytimes.com/2008/06/19/health/19iht-snbrody.1.1
    3772712.html
16. Based on the book of Michael Lewis (2003) about the success of
    manager Billy Beane and the Oakland Athletics
17. http://www.theguardian.com/football/blog/2015/feb/22/
    brentford-mathematical-modelling-denmark
18. https://twitter.com/NeelieKroesEU/status/442985759277346816,
    10 March 2014
19. https://www.lufthansagroup.com/en/responsibility/news/press-
    releases/singleview/archive/2013/march/19/article/2391.html
20. http://www.autoline.tv/daily/?p=38990
21. http://www.cnbc.com/id/38722872
22. http://www.kpmg.com/NL/nl/IssuesAndInsights/Arti
    clesPublications/Documents/PDF/Financial-Services/
    Frontiers-in-Finance-June-2015.pdf

2. **Remove Fear and Conquer Resistance**

1. http://www.nytimes.com/2012/02/19/magazine/shopping-
    habits.html
2. http://flowingdata.com/2012/02/16/companies-learn-
    your-secrets-with-data-about-you/
3. http://gizmodo.com/5679129/facebook-helps-predict-when-
    youre-likely-to-get-dumped

4. http://www.spiegel.de/international/zeitgeist/stephen-wolfram-analyses-personal-facebook-data-a-896981.html

5. http://doctorbeet.blogspot.co.uk/2013/11/lg-smart-tvs-logging-usb-filenames-and.html

6. http://www.technologyreview.com/news/524621/sell-your-personal-data-for-8-a-month/

7. http://www.usatoday.com/story/news/politics/2015/08/03/congress-pushes-rules-explosive-fertilizer/31075955/

8. http://dataprotectioneu.eu/index.html

9. http://www.zdnet.com/article/privacy-outrage-causes-bank-to-ditch-plans-for-targeted-ads-based-on-customers-spending-habits/

10. http://www.edelman.com/insights/intellectual-property/trust-2013/trust-across-sectors/trust-in-financial-services/

11. http://www.fastcompany.com/3027197/fast-feed/sorry-banks-millenials-hate-you

12. http://www.telegraph.co.uk/health/healthnews/10656893/Hospital-records-of-all-NHS-patients-sold-to-insurers.html

13. http://www.pcworld.com/article/240604/senator_calls_for_investigation_into_onstar_policy_changes_that_affect_customers_privacy.html

14. http://archive.wired.com/techbiz/it/magazine/16-03/ff_free?currentPage=all

15. Heineken was not the first to make this complaint. The American marketing pioneer, John Wanamaker, also made this complaint back in the nineteenth century. http://en.wikipedia.org/wiki/John_Wanamaker

16. http://infolab.stanford.edu/~backrub/google.html. Appendix A: Advertising and Mixed Motives

17. http://www.google.com/about/company/philosophy/

18. https://en.wikipedia.org/wiki/Cultural_relativism

19. Victor Lamme, De vrije wil bestaat niet, ('Free will does not exist'), Bert Bakker, 2010 (in Dutch) and Daniel M. Wegner, The Illusion of Conscious Will, 2002

20. http://www.nytimes.com/2013/07/28/magazine/should-reddit-be-blamed-for-the-spreading-of-a-smear.html
21. http://social.yourstory.com/2015/01/quotes-bill-gates-mobile-banking/
22. http://hbr.org/2011/01/the-big-idea-creating-shared-value
23. http://www.lifebuoy.com/socialmission/
24. https://sharedvalue.org/examples/lifebuoy-swasthya-chetna-soap-and-public-health-education
25. http://www.digitaltrends.com/mobile/mastercard-wants-link-smartphone-location-data-credit-card/
26. http://www.theguardian.com/technology/2011/apr/28/tomtom-satnav-data-police-speed-traps
27. http://www.theatlantic.com/technology/archive/2010/10/googles-ceo-the-laws-are-written-by-lobbyists/63908/

3. You Ain't Seen Nothing Yet

1   http://www.economist.com/news/briefing/21594264-previous-technological-innovation-has-always-delivered-more-long-run-employment-not-less
2. http://www.forbes.com/sites/pascalemmanuelgobry/2014/01/20/milton-friedman-predicted-the-rise-of-bitcoin-in-1999/
3. http://www.bloomberg.com/news/2013-12-04/greenspan-says-bitcoin-a-bubble-without-intrinsic-currencyvalue.html
4. http://krugman.blogs.nytimes.com/2013/12/28/bitcoin-is-evil/
5. A clear explanation in laymen's terms can be found on https://medium.com/p/73b4257ac833
6. http://guardianlv.com/2014/01/google-coin-may-stand-up-against-bitcoin/
7. http://uk.businessinsider.com/santander-has-20-25-use-cases-for-bitcoins-blockchain-technology-everyday-banking-2015-6
8. http://www.wired.com/2015/05/nasdaq-bringing-bitcoin-closer-stock-market/
9. www.ethereum.org, with a technical explanation of the philosophy on https://github.com/ethereum/wiki/wiki/White-Paper and a somewhat more accessible explanation on http://www.wired.com/2014/01/ethereum/

10. http://en.wikipedia.org/wiki/Robert_Solow

11. https://www.utwente.nl/en/news/!/2012/3/243583/dutch-waste-8-of-working-hours-on-it-problems-and-poor-digital-skills

12. http://www.economist.com/node/2052033

13. http://www.oxfordmartin.ox.ac.uk/downloads/academic/The_Future_of_Employment.pdf

14. http://www.wired.com/2012/12/ff-robots-will-take-our-jobs/

15. http://www.rethinkrobotics.com/products/baxter/

16. https://www.kickstarter.com/blog/introducing-launch-now-and-simplified-rules-0

17. http://ftalphaville.ft.com/2015/03/09/2120134/jobs-automation-engels-pause-and-the-limits-of-history/

18. http://www.calculatedriskblog.com/2014/06/may-employment-report-217000-jobs-63.html

19. http://www.imf.org/external/pubs/ft/fandd/2015/03/pdf/bessen.pdf

4.   Hitting the Bullseye First Time Around

1.   http://www.ncbi.nlm.nih.gov/pmc/articles/PMC3058895/

2.   http://gistsupport.medshelf.org/What_Milestone_in_GIST_Research_Happened_in_March_2000%3F

3.   http://onlinelibrary.wiley.com/doi/10.1111/joim.12325/full

4.   http://www.technologyreview.com/featuredstory/524531/why-illumina-is-no-1/

5.   http://www.wired.co.uk/news/archive/2014-01/15/1000-dollar-genome

6.   http://on.wsj.com/1PKvPXY

7.   http://www.wired.com/wiredscience/2014/02/elizabeth-holmes-theranos/

8.   http://engineering.columbia.edu/smartphone-finger-prick-15-minutes-diagnosis%E2%80%94done-0

9.   https://googleblog.blogspot.nl/2014/01/introducing-our-smart-contact-lens.html

10. Dutch Professor Bas Bloem demonstrated this in a TEDx Talk: https://www.youtube.com/watch?v=LnDWt10Maf8

11. http://www.express.co.uk/life-style/health/245664/How-a-humble-plaster-could-change-your-life
12. http://www.economist.com/node/21548493
13. http://venturebeat.com/2014/06/02/apple-announces-heath-kit-platform-and-health-app/
14. http://www.telegraph.co.uk/health/healthnews/10656893/Hospital-records-of-all-NHS-patients-sold-to-insurers.html
15. http://www.nytimes.com/2013/03/24/technology/big-data-and-a-renewed-debate-over-privacy.html
16. https://en.wikipedia.org/wiki/Watson_(computer)
17. http://www.techrepublic.com/article/ibm-watsons-impressive-healthcare-analytics-capabilities-continue-to-evolve/
18. http://www.wired.com/2014/08/enlitic/
19. http://techcrunch.com/2014/02/17/exogen-bio/
20. http://bits.blogs.nytimes.com/2014/02/04/outsourced-cancer-research-by-spacecraft/
21. http://www.forbes.com/sites/davidshaywitz/2014/07/04/google-co-founders-to-healthcare-were-just-not-that-into-you/

5. Data Stimuli for a Better World

1. http://www.nytimes.com/1990/12/25/health/what-if-they-closed-42d-street-and-nobody-noticed.html
2. http://en.wikipedia.org/wiki/Brundtland_Commission
3. http://assets.wnf.nl/downloads/wwf_livingplanetreport2010.pdf
4. https://en.wikipedia.org/wiki/Neoclassical_economics
5. The term 'tragedy of the commons' (as explored by Garrett Hardin, 1968) describes a situation where individuals acting independently and rationally according to each other's self-interest behave contrary to the best interests of the whole group by depleting some common resource. https://en.wikipedia.org/wiki/Tragedy_of_the_commons
6. An economy is Pareto efficient when every change in that economy results in an improvement of wealth for one, and at the same time loss of wealth for another. In other words, when one can only improve his or her position at the expense of another

7. http://www.nobelprize.org/nobel_prizes/economic-sciences/laureates/2007/press.html
8. http://www.jstor.org/discover/10.2307/29730088
9. http://www.de-gids.nl/artikel/humanisering-van-utopie-tot-architectuur
10. http://www.tudelft.nl/en/current/latest-news/article/detail/nieuw-regionaal-verkeersmodel-moet-files-in-de-stad-tegengaan/
11. http://www.telegraph.co.uk/technology/news/10544148/eBay-plans-gift-tokens-that-can-be-spent-at-any-shop-as-long-as-the-giver-approves.html
12. http://www.pnas.org/content/105/6/1786.full
13. http://www.oecd.org/greengrowth/48012345.pdf
14. https://www.kpmg.com/Global/en/IssuesAndInsights/ArticlesPublications/Documents/building-business-value.pdf
15. http://www.trucost.com/blog/105/the-true-cost-of-clothing
16. http://www.bloomberg.com/news/2014-05-01/how-accounting-stop-groaning-will-save-the-world.html
17. The motto of the MTL is 'Big Data Better Life'. You can find more information on the MTL at http://www.mobileterritoriallab.eu
18. Alex Pentland, 'Social Physics: How Good Ideas Spread', Penguin, 2014

6. Data Analytics Is the Society

1. http://www.ncbi.nlm.nih.gov/pubmed/15291423
2. Superintelligence: Paths, Dangers, Strategies (2014), Nick Bostrom
3. http://mindstalk.net/vinge/vinge-sing.html
4. Evgeny Morozov, The Net Delusion: The Dark Side of Internet Freedom, PublicAffairs, 2011
5. http://www.technologyreview.com/featuredstory/520426/the-real-privacy-problem/
6. https://www.utwente.nl/academischeplechtigheden/oraties/archief/2007-2014/oratieboekje_verbeek.pdf (in Dutch) and his associated book "Understanding and designing the morality of things" in English: http://press.uchicago.edu/ucp/books/book/chicago/M/bo11309162.html

7. Mayer-Schönberger also recognizes the dangers of Big Data and the subtitle of his book 'Big Data: A Revolution That Will Transform How We Live, Work, and Think' quoted here is therefore not entirely representative of his ideas. He argues rather for a responsible use of Big Data in order to avoid dystopias

8. http://www.telegraph.co.uk/education/7585505/Teaching-Inspiring-British-children-Slumdog-style.html

9. https://www.theinformation.com/Google-beat-Facebook-For-DeepMind-Creates-Ethics-Board

10. http://www.theguardian.com/science/2015/may/21/google-a-step-closer-to-developing-machines-with-human-like-intelligence

7. Wanted: Thousands of Sherlock Holmes Clones

1. http://tylervigen.com/spurious-correlations

2. http://sherlockholmesquotes.com/

3. http://online.wsj.com/articles/SB124967937642715417

4. http://www.economist.com/blogs/newsbook/2010/10/what_caused_flash_crash

5. http://www.wired.com/2013/02/big-data-means-big-errors-people/

6. It cannot be a coincidence: Watson is the name of Sherlock Holmes' assistant, and it is also the name of the supercomputer developed by IBM—named after IBM's first CEO

7. http://www.zerohedge.com/news/2014-04-28/elephant-room-deutsche-banks-75-trillion-derivatives-20-times-greater-german-gdp

8. http://www.reuters.com/article/deutsche-bank-capital-ecb-idUSL6N0NL28C20140429

9. http://www.independent.co.uk/life-style/gadgets-and-tech/news/google-chief-my-fears-for-generation-facebook-2055390.html

8. The Question Is More Important Than the Answer

1. https://en.wikipedia.org/wiki/Socratic_method

2. http://www.nytimes.com/2003/05/11/national/11PAPE.html

3. The Filter Bubble: What the Internet Is Hiding from You, Eli Pariser, 2011
4. http://www.google.com/transparencyreport/removals/europeprivacy/
5. The Facebook Effect, David Kirkpatrick, Simon & Schuster, 2011
6. http://www.technologyreview.com/view/522111/how-to-burst-the-filter-bubble-that-protects-us-from-opposing-views/
7. http://www.hbo2025.nl/wp-content/uploads/2014/05/Essay-prof-dr-Dirk-Van-Damme.pdf in Dutch, derived from the essay http://www.ippr.org/files/images/media/files/publication/2013/04/avalanche-is-coming_Mar2013_10432.pdf
8. It is a difficult to translate this concept, attributed to the German scientist Wilhelm von Humboldt. Self-development and general education comes closest. Bildung represents both a humanist and a political ideal. The 'Bildung' that von Humboldt strives for is a general development of all human qualities. https://en.wikipedia.org/wiki/Bildung
9. https://en.wikiquote.org/wiki/Eric_Hoffer
10. http://www.rijksoverheid.nl/documenten-en-publicaties/toespraken/2014/03/17/lezing-minister-bussemaker-bij-symposium-knaw-vaardigheden-voor-de-toekomst.html (in Dutch)
11. Adapted from http://www.ed4wb.org/?p=778

9. Transparency, a Weapon for Citizens

1. http://www.airspacemag.com/history-of-flight/show-me-the-way-to-go-home-9723795/
2. https://wikileaks.org/About.html
3. https://en.wikipedia.org/wiki/Kony_2012
4. Mind over matter: why intellectual capital is the chief source of wealth, Ronald J. Baker
5. http://www.wired.com/wired/archive/15.04/wired40_ceo.html
6. http://www.theguardian.com/politics/blog/2009/jun/19/mps-expenses-what-you-ve-found
7. http://ec.europa.eu/archives/commission_2010-2014/kroes/en/blog/opendata.html
8. http://www.ted.com/talks/tim_berners_lee_the_year_open_data_went_worldwide

9. http://www.iamexpat.nl/read-and-discuss/expat-page/news/dutch-hospital-patient-figures-difference-treatment

10. See also research from David Dranove, Daniel Kessler, Mark McClellan and Mark Satterthwaite in 2002. http://www.nber.org/papers/w8697

11. https://www.ted.com/talks/hans_rosling_shows_the_best_stats_you_ve_ever_seen

12. http://www.nyasatimes.com/2013/12/27/report-says-malawi-among-best-performing-economies-jb-upbeat/

13. http://www.ft.com/intl/cms/s/2/785bd614-9378-11e3-b07c-00144feab7de.html#axzz2tKPx7zq0

14. http://www.economist.com/news/business/21645745-management-goal-setting-making-comeback-its-flaws-supposedly-fixed-quantified-serf

10. Systems Determine Our Behavior

1. http://www.nytimes.com/2008/02/17/nyregion/nyregionspecial2/17colwe.html

2. https://en.wikipedia.org/wiki/Samuel_J._Tilden

3. http://en.wikipedia.org/wiki/Block_booking

4. http://www.dw.com/en/german-minister-google-breakup-may-be-required/a-17641881

5. http://www.reuters.com/article/us-google-antitrust-idUSKCN0J525V20141121

6. http://www.americanbar.org/content/dam/aba/publishing/antitrust_source/apr15_full_source.authcheckdam.pdf

7. http://en.wikipedia.org/wiki/A_Declaration_of_the_Independence_of_Cyberspace

8. http://reason.com/archives/2004/08/01/john-perry-barlow-20

9. http://www.wired.co.uk/news/archive/2015-10/27/net-neutrality-european-union-vote

10. https://www.youtube.com/watch?feature=player_embedded&v=fpbOEoRrHyU

11. **A New Ecosystem**

1. European Data Grid (edg), Enabling Grids for E-Science in Europe (egee), European Grid Infrastructure (egi)
2. http://www.popmednet.org/?page_id=45#PMN_is
3. https://datascience.berkeley.edu/anonymous-data/
4. http://www.journalofaccountancy.com/issues/2000/feb/whoare weasaprofessionandwhatmustwebecome.html
5. http://www.fastcompany.com/1249710/freerisk-crowdsourcing-credit-ratings
6. http://www.teslamotors.com/blog/all-our-patent-are-belong-you. For the nerds among us, this is a reference to the famous 4Chan meme 'all your base are belong to us'; see http://en.wikipedia.org/wiki/All_your_base_are_belong_to_us.
7. http://www.slate.com/articles/technology/history_of_innovation/2014/05/william_gilmour_power_loom_the_industrial_revolution_and_open_innovation.html
8. These three issues are discussed by Sangeet Paul Choudary in an article in the Harvard Business Review. See http://blogs.hbr.org/2013/01/three-elements-of-a-successful-platform/
9. http://www.nytimes.com/2013/02/19/opinion/brooks-what-data-cant-do.html

Epilogue

1. http://www.zdnet.com/article/gmail-is-boiling-the-frog-and-we-are-the-frog/

Index

© Atlantis Press and the author(s) 2016
S. Klous and N. Wielaard, *We are Big Data*,
DOI 10.2991/978-94-6239-183-3

CPSIA information can be obtained
at www.ICGtesting.com
Printed in the USA
LVOW01*1710310317

529199LV00012B/177/P